Stars and Galaxies

Stars and Galaxies
ASTRONOMY LIBRARY NO. 4

ASTRONOMY's Guide to Observing the Cosmos

David J. Eicher, editor
Associate Editor, ASTRONOMY

With contributions by Alan Goldstein, Phil Harrington, David Higgins, Richard W. Jakiel, Alister Ling, Steve Lucas, Rod Pommier, Max Radloff, Chris Schur, and Gregg D. Thompson

AstroMedia
Waukesha, Wisconsin
1992

Books by David J. Eicher

The Universe from Your Backyard; a guide to deep-sky objects from ASTRONOMY magazine (Cambridge University Press, New York, and Kalmbach Publishing Co., Waukesha, Wisconsin, 1988)

Deep Sky Observing with Small Telescopes; a guide and reference (Editor and coauthor; Enslow Publishers, Hillside, New Jersey, 1989)

Beyond the Solar System; 100 best deep-sky objects for amateur astronomers (AstroMedia Corp., Waukesha, Wisconsin, 1992)

Stars and Galaxies; ASTRONOMY's guide to exploring the cosmos (Editor and coauthor; AstroMedia Corp., Waukesha, Wisconsin, 1992)

FOR ROBERT BURNHAM, whose considerate mind is expert not only in astronomy but across the whole spectrum of life.

Published by AstroMedia, a Division of Kalmbach Publishing Co., 21027 Crossroads Circle, Waukesha, WI 53187.

© 1992 Kalmbach Publishing Co.
All rights reserved.
Printed in Singapore.

Editorial staff: Richard Berry, David Bruning, Robert Burnham, Stephen Cole, Alan Dyer, Michael Emmerich, Robert Hayden Jr., Jeff Kanipe, Kristine R. Majdacic, Christine Reel, Rhoda Sherwood, Tracy Staedter, Margaret Sullivan, and Richard Talcott.

Art and Production staff: Thomas L. Hunt (art director), Larry Luser (production coordinator), Todd Graveline, Patti L. Keipe, Lisa Bergman.

Library of Congress Cataloging-in-Publication Data

Stars and galaxies :   a reprint from Astronomy magazine / [compiled] by
   David J. Eicher.
      p.   cm.
   Includes bibliographical references and index.
   ISBN 0-913135-05-4
   1. Stars. 2. Galaxies. I. Eicher, David J., 1961-
II. Astronomy.
QB801.S725   1992
523.8--dc20
                                                                91-45407
                                                                   CIP

Contents

Preface by David J. Eicher	7

General Deep-Sky Observing

Treasures of the Winter Milky Way by David J. Eicher	8
A New Dimension to Winter Deep-Sky Objects by David J. Eicher	16
Journey to the Center of the Galaxy by Phil Harrington	22
Magnificent Orion by Alan Goldstein	28
The Wonders of the Coma Cluster by Alan Goldstein	32
Explore the Southern Sky by Gregg D. Thompson	36

Galaxies

Observing the Andromeda Galaxy by Alan Goldstein	44
Galaxy Hunting around the Big Dipper by Alan Goldstein	50
Legendary Faint Galaxies by Steve Lucas	58
The Galaxies of Cetus by David Higgins	62
Explore the Virgo Cluster by Alan Goldstein	68
Observing the Sculptor Group of Galaxies by Rod Pommier	76
The Galaxies of Sextans by David Higgins	80
Galaxy Hunting in the Great Bear by Richard W. Jakiel	84
Galaxies of the Great Square by David Higgins	88
Autumn's Galaxies: The Best and Brightest by Max Radloff	94

Stars and Clusters

Standout Winter Star Clusters by Alister Ling	102
Dramatically Diverse Globulars by Chris Schur	108
Scanning the Scutum Starcloud by Phil Harrington	114
Splashy Summer Star Clusters by Alan Goldstein	118
Southern Clusters of All Ages by Gregg D. Thompson	121
Exploring Open Clusters in Canis Major by Phil Harrington	126
The Distant Suns by David Higgins	130
Split a Star in Two by Alan Goldstein	134
Five Challenging Globulars by David Higgins	138

Bright and Dark Nebulae

The Secret World of Dark Nebulae by David Higgins and David J. Eicher	142
The Art of Observing Planetaries by David J. Eicher	148
The Ghostly Glow of Gaseous Nebulae by David Higgins	151
The Challenge of Dusty Dark Nebulae by David Higgins	154
Tracking down the Helix by David Higgins	157
Stalking the Elusive Horsehead by David Higgins	160
Find a Supernova Remnant by Chris Schur	166
The Challenge of Winter Nebulae by Phil Harrington	172
Observing Bright Planetary Nebulae by Alan Goldstein	176
Observing Nebulosities in Cygnus by Alan Goldstein	182
Great Summer Planetaries by Alan Goldstein	188

Bibliography	193
The Authors	196
Index	198

Preface

Stars and galaxies are for everyone. Unlike many sciences, astronomy is a people's science in that anyone beneath a clear sky can participate in it. With a simple pair of binoculars or a small telescope, you can visit the cosmos from your back yard. All you need is information on what to look at and how to interpret what you're seeing. *Stars and Galaxies for Everyone* is intended to introduce you to several hundred of the brightest and most beautiful — and many of the most challenging — citizens of our Milky Way Galaxy and other galaxies beyond.

The book contains articles reprinted from ASTRONOMY magazine. After publishing our first book of deep-sky object reprints, *The Universe from Your Backyard* (Cambridge University Press and Kalmbach Publishing Co., 1988), we began a new series of deep-sky observing articles in ASTRONOMY. While *The Universe from Your Backyard* took you on a constellation-by-constellation tour of the sky, *Stars and Galaxies* is focused on richly interesting regions of sky. Over the course of more than three years — between the September 1988 issue and the December 1991 issue — we printed 37 articles with practical information on deep-sky objects in these regions and how to observe them. The articles alternated between describing regions containing bright objects easily visible in small telescopes and introducing fainter objects challenging to more experienced observers.

I hope that you'll come to use this work both beside your telescope and as a reference on your astronomy book shelf. I hope that whether you are a beginner in astronomy or an old hand, you'll allow this book to introduce you to a galaxy here and a nebula there, objects that will become old friends after several seasons of observing.

A word about the data in this book: statistical information on deep-sky objects is a field in its infancy. In many observing guides of the 1970s and 1980s, much of the published information on, say, a galaxy's brightness, size, distance, and so on, was sadly outdated. In this work we have drawn upon two principal sources in an effort to use the best data we could — *Observing Handbook and Catalogue of Deep-Sky Objects* by Christian Luginbuhl and Brian A. Skiff (Cambridge University Press, New York, 1990) and *Sky Catalogue 2000.0,* volume 2 (Cambridge University Press and Sky Publishing Corp., New York, 1985). These authoritative works contain information of the highest reliability.

Unlike the articles making up *The Universe from Your Backyard*, the articles in this book were written by a diverse group of amateur deep-sky observers. They are Alan Goldstein, Earth and natural sciences coordinator at Louisville's Museum of Science and History; Phil Harrington, an engineer and science educator living in Smithtown, New York; David Higgins, a sergeant in the U.S. Army Reserve; Richard W. Jakiel, an Atlanta-based graduate student studying astronomy; Alister Ling, a meteorologist in Canada; Steve Lucas, a Chicagoan interested in supernovae; Rod Pommier, a a surgical oncologist at the Oregon Health Sciences University in Portland; Max Radloff, a musician and talented astronomer from St. Paul, Minnesota; Chris Schur, an engineer and astrophotographer living in Payson, Arizona; and Gregg D. Thompson, an artist, designer, and deep-sky sketcher based in Queensland, Australia. Their tireless energy under the stars, expertise at the eyepiece, and keen eye for detail when viewing distant galaxies, is a fitting tribute to the accomplishments of backyard astronomers everywhere.

David J. Eicher
Waukesha, Wisconsin
March 1992

The Double Cluster
Magnitude 4.3p and 4.4p   Sizes 30′

Treasures of the Winter Milky Way

Perseus and Cassiopeia in the northern sky offer
a bounty of bright objects for observers with nothing more than binoculars.
by David J. Eicher

Every cold, crisp autumn night that pulls me out under the stars attracts me to one special place in the sky. It's an area densely stocked with impressive deep-sky objects. The region of sky encompasses the constellations Perseus and Cassiopeia and represents the richest region of the winter Milky Way. When we look toward Perseus and Cassiopeia, we are gazing through an outer spiral arm, the Perseus arm, toward the edge of our Galaxy's disk. We're looking into the Galaxy's plane of rotation — the space that we are slowly rotating toward. I start an annual reconnoitering of this wondrous area by simply gazing up at the Double Cluster in Perseus, a double patch of hazy light to the naked eye. The mystery and sense of remoteness inspired by the Double Cluster soon has me operating at full tilt with my telescope.

The **Double Cluster** in Perseus serves as a good object for testing a night's transparency. Composed of two large and bright open clusters, the Double Cluster is one of the brightest deep-sky objects in the sky. From a reasonably dark suburban or rural site, the Double Cluster is immediately obvious to the naked eye, appearing as a fuzzy barbell between the constellations of Perseus and Cassiopeia. If the cluster is not easily visible, the sky is not transparent enough for good observing.

Telescopes in the 2-inch to 6-inch range show the Double Cluster's western component (NGC 869) and eastern component (NGC 884) as roughly equivalent: each is a sparkling disk of tiny blue-white stars nearly as large as the Full Moon. Viewing the Double Cluster (and other big open clusters) with low power generally provides the most pleasing scene. Clusters appear as clusters when magnifications are low enough to surround them with empty sky. If you magnify them too much, clusters appear to be merely rich backgrounds of stars evenly spread across the field of view. A magnification between 20x to 40x is best for viewing big open clusters.

In large telescopes the Double Cluster comes alive. Each cluster contains at least 150 stars arranged in haphazard patterns — curving chains, clumps, pairs, triangles, and straight lines. (Modern studies identify two hundred stars in NGC 869 and 150 in NGC 884, although each cluster undoubtedly contains additional members.) The brightest stars in the clusters glow at magnitudes 6.6 (NGC 869) and 8.1 (NGC 884). Most of the brightest stars appear pure white or bluish white in color. However, a noticeable few have a yellowish or golden orange tinge. These varied colors help to give the Double Cluster a jewel-like quality.

After poring over the Double Cluster, nudge your telescope about 4° north and 1° east and you'll come upon a big, bright cluster in Cassiopeia, IC **1805** (Melotte 15). Often overlooked because of its exclusion from the Messier and NGC catalogs, IC 1805 is nonetheless quite an impressive object. Consisting of forty stars tucked into a diameter of 22′, this cluster is an easy object for binoculars and small scopes. The brightest individual star glows at magnitude 7.8, and the group's overall magnitude is 6.5 — at the limit of naked-eye visibility from an exceptionally dark site. If you sweep

The California Nebula
Magnitude ★ 4.0   Size 145′ by 40′

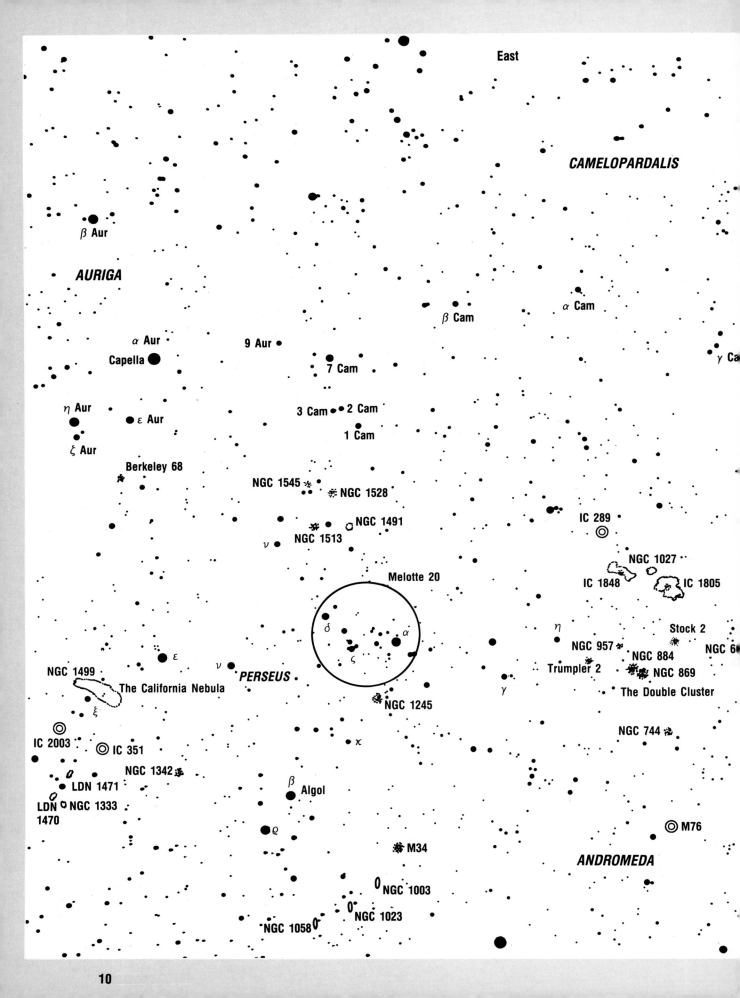

Deep-Sky Objects in Perseus and Cassiopeia

Object	Type	R.A. (2000.0) Dec.	Mag.	Size	N★
Berkeley 58	oc	0h00.2m +60°58'	9.7	8'	30
IC 10	eg	0h20.4m +59°18'	10.3	5.1'x4.3'	—
NGC 103	oc	0h25.3m +61°21'	9.8	5'	30
NGC 129	oc	0h29.9m +60°14'	6.5	21'	150
NGC 136	oc	0h31.5m +61°32'	—	1.2'	20
NGC 147	eg	0h33.2m +48°30'	9.3	12.9'x8.1'	—
IC 63	ern	0h59.5m +60°49'	—	10'x3'	2.5v
NGC 457	oc	1h19.1m +58°20'	6.4	13'	80
NGC 559	oc	1h29.5m +63°18'	9.5	4.4'	60
M103	oc	1h33.2m +60°42'	7.4	6'	25
M76	pn	1h42.4m +51°34'	12.2p	65"	17.0
NGC 663	oc	1h46.0m +61°15'	7.1	16'	80
NGC 744	oc	1h58.4m +55°29'	7.9	11'	20
Stock 2	oc	2h15.0m +59°16'	4.4	60'	50
NGC 869	oc	2h19.0m +57°09'	4.3p	30'	200
NGC 884	oc	2h22.4m +57°07'	4.4p	30'	150
IC 1805	en	2h33.4m +61°24'	—	60'x60'	—
NGC 957	oc	2h33.6m +57°32'	7.6	11'	30
Trumpler 2	oc	2h37.3m +55°59'	5.9	20'	20
NGC 1003	eg	2h39.3m +30°09'	11.5	5.4'x2.1'	—
NGC 1023	eg	2h40.4m +39°04'	9.5	8.7'x3.3'	—
M34	oc	2h42.0m +42°47'	5.2	35'	60
NGC 1027	oc	2h42.7m +61°33'	6.7	20'	40
NGC 1058	eg	2h43.5m +37°21'	11.5	3.0'x2.9'	—
IC 1848	en	2h51.3m +60°25'	—	60'x30'	—
IC 289	pn	3h10.3m +61°19'	12.3p	34"	15.9
NGC 1245	oc	3h14.7m +47°15'	8.4	10'	200
Melotte 20	oc	3h22.m +49°	1.2	185'	50
NGC 1333	rn	3h29.3m +31°25'	—	9'x7'	9.5
NGC 1342	oc	3h31.6m +37°20'	6.7	14'	40
LDN 1470	dn	3h44.0m +31°47'	—	100'x70'	—
IC 351	pn	3h47.5m +35°03'	12.4p	7"	14.8
LDN 1471	dn	3h48.0m +32°54'	—	22'x9'	—
IC 2003	pn	3h56.4m +33°52'	12.6p	7"	16.5v
NGC 1499	en	4h00.7m +36°37'	—	145'x40'	4.0
NGC 1491	en	4h03.4m +51°19'	—	3'x3'	11
NGC 1513	oc	4h10.0m +49°31'	8.4	9'	50
NGC 1528	oc	4h15.4m +51°14'	6.4	24'	40
NGC 1545	oc	4h20.9m +50°15'	6.2	18'	20
NGC 1579	rn	4h30.2m +35°16'	—	12'x8'	12

Object: Designation in Messier, NGC, or other catalog. **R.A. and Dec.:** Right ascension and declination in equinox 2000.0. **Mag.:** V magnitude of object unless followed by a p (photographic magnitude). **Size:** Diameter in arcminutes or arcseconds. **N★:** Number of stars in cluster, brightness of illuminating star (nebulae), or brightness of central star (planetary nebulae). V following magnitude denotes variability.

Open Cluster NGC 457
Magnitude 6.4   Size 13'

around IC 1805 at low power using a telescope of 8-inches or larger aperture, you may see traces of very faint, soft nebular light. This is light from an emission nebula surrounding the cluster that measures 60' across. Like the material in the Orion Nebula, this material is coalescing to form the stars in the central cluster. Although this nebula is difficult to observe, it can be photographed on red-sensitive film with relative ease.

Just 2° southeast of IC 1805 lies another giant complex of star cluster and emission nebula, **IC 1848**. The star cluster IC 1848 contains ten stars of magnitude 7.1 and fainter in an area only 12' across. It is an attractive specimen for binoculars and small telescopes. The emission nebula IC 1848 is shaped like a squashed rectangle spanning 60' by 30'. The nebula's surface brightness is rather low, making it visible only with averted vision in an 8-inch telescope — and then only on very dark nights. IC 1805 and IC 1848 offer a rare comparison of two similar star-forming regions located at approximately the same distance (2.2 kiloparsecs) and positioned near each other in the sky. IC 1848 is fainter and has only a small star cluster because it is younger than IC 1805 and hasn't had as much time to produce new stars.

After viewing IC 1805 and IC 1848, drop back to the Double Cluster and continue moving your scope southwest. You'll encounter the 5th-magnitude star 4 Persei and then Phi (φ) Persei, a 4th-magnitude bluish white variable star. A move 1° north of this star will center your telescope on one of the most beautiful planetary nebulae in the sky, **M76** (NGC 650-1).

Often called the Little Dumbbell Nebula because of its resemblance to M27 in Vulpecula, M76 is revered by many observers as one of the faintest and most challenging Messier objects, a reputation it well deserves. M76 has the two important characteristics that make deep-sky objects challenging: a large diameter (65") and a dim magnitude (photographic magnitude 12.2). Together, these give M76 a low surface brightness that makes the object appear dim no matter how much aperture you throw at it. To ensure a good view of M76, wait for a dark, moonless night with steady seeing and excellent transparency.

As you inch your telescope up toward the familiar W-shaped pattern of Cassiopeia, you'll encounter many bright deep-sky objects. Two small, attractive star clusters lie between Epsilon (ε) and Delta (δ) Cassiopeiae, the two easternmost stars in the W. **M103** (NGC 581) is a smattering of faint stars held within a diameter of just 6'. Altogether, twenty-five stars of magnitude 10.6 and fainter form this group. Two degrees east and a few arc-minutes north of M103 lies a larger, brighter cluster. **NGC 663** spans 16' and holds eighty stars of magnitude 8.4 and fainter within its gravitational grasp. The total magnitude of NGC 663 is 7.1; M103's is 7.4.

The rich cluster **NGC 559** lies 3° north of M103. This unusual cluster has an overall magnitude of 9.5 and is comprised of light from sixty stars glowing at magnitude 10.6 and fainter. All of these faint stars are packed into a diameter of 4.4', making this group appear as luminescent buckshot punched through the dark background of the sky. As you compare these clusters, consider how the concentrations, colors, and brightnesses of the stars vary. Do the clusters appear differently at different magnifications?

One of the most unappreciated open clusters, **NGC 457**, lies 3° south of Delta Cassiopeiae, nearly directly

The Bubble Nebula
Magnitude ★ 6.9   Size 15′ by 8′

Emission Nebula NGC 281
Magnitude ★ 7.8   Size 35′ by 30′

The Little Dumbbell Nebula
Magnitude 12.2p   Size 65″

Nebulae IC 59 and IC 63
Magnitude★ 2.5v   Sizes 10′ by 5′ and 10′ by 3′

behind the ruddy, 5th-magnitude star Phi (φ) Cassiopeiae. The bright star Phi and a slightly dimmer star form what look like owl eyes, and the pattern of a few dozen bright stars that make up the torso, wings, and legs is striking. NGC 457's unmistakable shape takes me back to a night in 1979 when I looked at this cluster and the pattern of stars was so distinctive that I began calling it the "Owl Cluster." Some eighty stars of magnitude 8.4 and fainter make up the Owl Cluster, giving it a total magnitude of 6.4 spread over 13′. This object is beautiful in binoculars and small telescopes and requires a wide field with large telescopes. Astronomers debate about whether or not Phi Cassiopeiae is a physical member of the group. If it is, because of the cluster's distance of 2.8 kiloparsecs, it is one of the most luminous stars known.

The erratic variable star Gamma (γ) Cassiopeiae, the central star in the W, plays host to two exceptionally low surface brightness nebulae that shine by the bright star's reflected light. **IC 59** and **IC 63** are pockets of dust lying near Gamma in space, probably composed of particles about the size of those in cigarette smoke. Rather than shining by fluorescence like an emission nebula, these objects simply bounce light from Gamma toward our line of sight. They would be invisible if Gamma didn't lie so close — about 30 parsecs away.

Moving south from Gamma, aim your telescope toward Eta (η) Cassiopeiae, a 4th-magnitude double star. From there, head south 2° and you'll see a triangle of 7th-magnitude stars, two of which are doubles. Surrounding this triangle is the elusive emission nebula **NGC 281**, a low surface brightness object that's challenging in any telescope. A wide field of view, a dark night, and a nebular filter are essential for spotting this

Galaxy NGC 1023
Magnitude 9.5 Size 8.7' by 3.3'

Galaxy NGC 278
Magnitude 10.9 Size 2.2' by 2.1'

hazy cloud. Look for filaments of gray-green light measuring a total of 35' by 30' in extent.

Six degrees northwest of Eta lies Beta (β) Cassiopeiae, the western end of the W. Positioned about 2° east of this bright star is **IC 10**, a galaxy that belongs to our Local Group and barely shines through the obscuring disk of Milky Way stars, gas, and dust. Large telescopes are required to spot this galaxy. On an exceptional night an 8-incher will show IC 10 as a fuzzy smear some 2' across. But to see a central condensation and the galaxy's full extent (5.1' by 4.3'), you'll need a 16-inch or larger telescope.

Now center your scope back on Beta. Roughly 3° southwest of this star lies one of the richest open clusters in the sky, **NGC 7789**. This cluster is easy to locate because it is placed midway between the 4th- and 5th-magnitude stars Rho (ϱ) and Sigma (σ) Cassiopeiae. Some

three hundred stars of magnitude 10.7 and fainter crowd a diameter of 16' in this impressive cluster. Visible as a ghostly glow in binoculars, the group appears as a rich swarm of white points of light in telescopes.

The northwestern reaches of Cassiopeia play home to two strikingly unusual objects. **M52** (NGC 7654), a bright, rich open cluster, and **NGC 7635**, an odd emission nebula, lie within a degree of each other. M52 is a stunning collection of one hundred 8th-magnitude and fainter stars packed into a diameter of 13', giving the group a total magnitude of 6.9. The most notable characteristic of M52 is its great range of star colors, from white and blue-white to burnt orange. NGC 7635 lies 1° southwest of M52 and is nicknamed the Bubble Nebula because of the wispy shell structure revealed in long-exposure photographs. This nebula is visible in a 6-inch telescope, although it is quite faint due to low surface brightness.

Back in Perseus, south of the Double Cluster, lies a line of bright deep-sky objects accessible in small telescopes. **M34** (NGC 1039) is a striking open cluster composed of sixty white stars in a diameter as large as the Full Moon. Perfect for binocular observers, this cluster should by viewed at low powers when telescopes are employed. Several degrees south of this cluster lie three galaxies aligned north-south. **NGC 1003**, the northernmost galaxy, is a magnitude-11.5 spiral measuring 5.4' by 2.1'. Two degrees south lies **NGC 1023**, the finest galaxy in Perseus. This object glows at magnitude 9.5 and spans 8.7' by 3.3'. It is an E7 elliptical galaxy viewed nearly edge-on to our line of sight. **NGC 1058** is a magnitude-11.5 spiral that appears almost exactly round in telescopes.

The most mysterious and challenging deep-sky object in Perseus is **NGC 1499**, popularly known as the California Nebula due to its distinctive shape. An enormous but extremely faint emission nebula, the California is visible only under the best conditions — on dark nights when the contrast between object and sky is at its best. You'll probably need a 10-inch or larger scope to see it, although some observers report spotting the nebula's brightest parts with telescopes as small as 4 inches in aperture.

To find the California Nebula, sweep slowly north of the 4th-magnitude star Xi (ξ) Persei, which illuminates the nebula. Use wide-field binoculars or an extremely wide-field, low-power eyepiece. If you observe on an exceptionally dark night, you may see bright parts of the nebula's light as wispy, filamentary bands of faint light. In some areas these bands of light appear linear; in other parts they are strongly curved. If you observe the California Nebula, make a sketch or at least take notes describing what you see. You'll be in a relatively small group of people who have "conquered the California."

Whether you watch the bright clusters or stalk the faint nebulae, the region of sky containing Perseus and Cassiopeia offers a grand selection of targets for your telescope. Use the observing opportunities you have this autumn to get out under the stars and challenge yourself to experiment with these objects. View them at low powers, at high powers, with nebular filters, and with averted vision. Challenge yourself to see details that have previously escaped you. The treasures of Perseus and Cassiopeia, crown of the winter Milky Way, offer not only interesting things to see but also a fresh opportunity to sharpen your observing abilities. ☐

A New Dimension to Winter Sky Objects

Photographs reveal the faint structure, the breathtaking colors, and the full extent of winter deep-sky wonders.
by David J. Eicher

Astrophotography is powerful stuff. It can extend your visual observing experiences by providing a new dimension to what you see in a telescope. Subtle details, colors, and overall structures too faint to be seen with the eye come alive in long-exposure photographs. By doing a second type of observing — carefully studying photographs — you can see the sky in a new light.

Film provides a fresh look at deep-sky objects in two ways. First, film can store up light during a long exposure. Almost magically this lets a bright image emerge after many minutes of collecting faint light. Second, film records color in faint objects because it retains a broad spectral sensitivity at low light levels. The human eye cannot do this because your retina's color receptors respond only to relatively bright light. When it comes to "seeing" faint objects and detecting color in them, film gives you an important, different look at your favorite objects.

The winter Milky Way contains intricately detailed deep-sky wonders. One of the most beautiful is the Rosette Nebula in Monoceros, a cloud of gas surrounding the bright star cluster NGC 2244. The stars in the cluster are forming out of the surrounding gas, which makes the Rosette one of the nearest and grandest star-making factories known in our Galaxy.

The eye cannot do justice to the Rosette Nebula: only photography reveals its full glory. Small scopes show the star cluster but not the nebula. Large scopes reveal faint traces of the nebula but show no details to speak of. Long-exposure photos, however, reveal the Rosette in all its splendor.

The photo on the opposite page shows the Rosette's overall structure and network of fine detail. Dozens of bright and faint cluster stars lie within the nebula's central dark "hole," a void created by the strong winds produced by the cluster's stars. Bright ridges of glowing nebulosity curve around the Rosette's center and backlight tiny, dark specks called Bok globules. (These are clumps of dust slowly condensing into stars.) This photograph transports you far deeper into the Rosette than any telescope can and lets you see the object as a complex, active region with many distinct features.

West of the Rosette Nebula lies Orion the Hunter. A small, inconspicuous patch of light tucked in the Hunter's sword is the Orion Nebula, M42. The Orion Nebula is the brightest star-forming region visible from the Northern Hemisphere.

The eye shows the Orion Nebula as a pale, green, misty cloud with several stars embedded in it. Long-exposure photographs reveal that the entire sword of Orion is covered with nebulosity. The top photo on page 97 shows the Orion Nebula lost in a blanket of dust and gas. Faint streamers of nebulosity extend away from the white, overexposed Orion Nebula. The difference between the nebula in a telescope and its shape in a "deep" photo of the region is breathtaking.

A long-exposure photograph reveals that Orion is draped in swaths of nebulosity. The top photo on page 96 shows a veil of glowing light spread across the constellation, as well as five "hot spots" where nebulae are visible. A ruddy glow lies below the belt stars in Orion. A group of nebulae lies in this area, including NGC 2024, IC 434, and the Horsehead Nebula.

Forming a giant arc in the lower half of the photograph is Barnard's Loop, a bubble of warm hydrogen gas expanding in Orion's Milky Way. The brightest pieces of this object are visible in large backyard telescopes, but photography enables us to see Barnard's Loop as a single object. In the loop, bright areas twist like puffs of smoke. These mark the shock front where warm gas inside the bubble is colliding with the cold material in the surrounding space and causing the nebulosity to shine.

At the top of Orion is a smaller, circular glow called the Lambda Orionis Nebula. This is also a bubble of gas. However, it is younger than Barnard's Loop and appears less spread out. This object is not visible in telescopes: photography alone reveals its existence.

In the vicinity of Zeta Orionis, the easternmost belt star, a glow cuts by a dark patch east of Zeta. Large telescopes show a faint, gray spot of nebulosity southeast of the bright star. The largest backyard scopes, used on a perfect night, reveal the Horsehead Nebula south of Zeta.

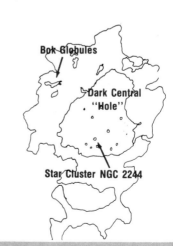

A WREATH OF SOFT LIGHT, the Rosette Nebula in Monoceros shows tiny dark globules in photographs. Only the central star cluster is visible in small telescopes. Photo by Tony Hallas and Daphne Mount.

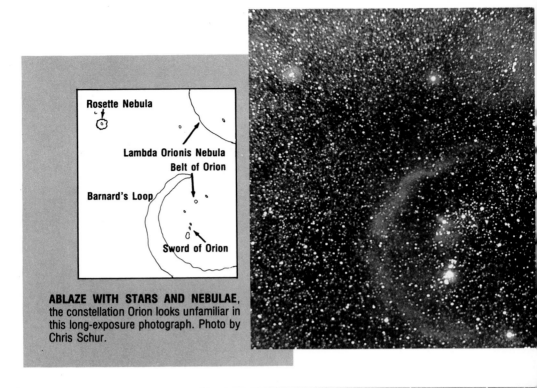

ABLAZE WITH STARS AND NEBULAE, the constellation Orion looks unfamiliar in this long-exposure photograph. Photo by Chris Schur.

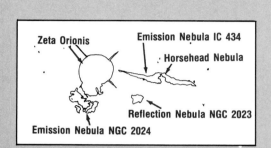

COLORS INVISIBLE TO THE EYE come alive in photographs. Jack Child's photo of the Horsehead Nebula region in Orion shows tawny yellow NGC 2024, reflection nebula NGC 2023, and the strip of light, IC 434, that lies behind the Horsehead.

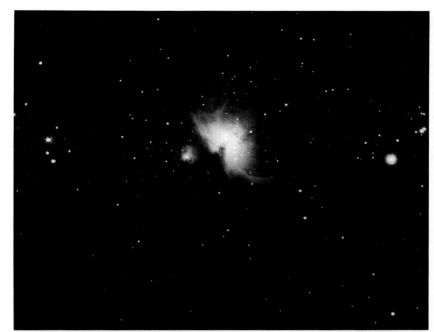

THE VISIBLE ORION NEBULA, simulated by a short-exposure photograph (top), is a roughly circular patch of greenish light 1° across. Photo by Dan Gordon. More exposure time reveals that the entire sword of Orion (above) is engulfed in nebulosity, far too faint to be observed with a telescope. Photo by Robert Reeves.

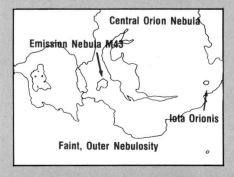

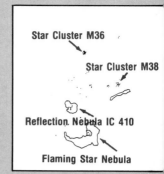

OBJECTS JUMP OUT when photographed in the right color of light. Central Auriga in blue light (below left) shows star clusters, but the same region in red light (below right) shows an extensive network of nebulosity out of the reach of observers. Photos by Ronald Royer.

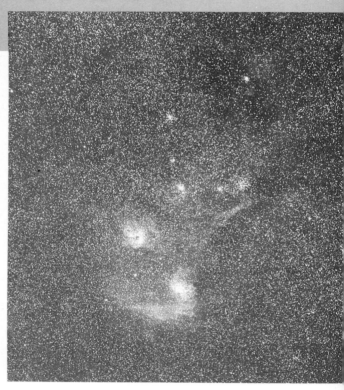

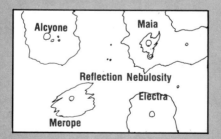

WISPS OF BLUE GAS surround bright young stars in the Pleiades star cluster in Taurus. Visible in telescopes only with great difficulty, photography shows the Pleiades nebulosity in a 5-minute exposure. Photo by Jack Newton.

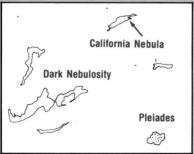

THE CALIFORNIA NEBULA in Perseus is a difficult target for telescopes, faintly visible on the best nights. Photography shows its outline and the surrounding lanes of dark nebulae. Photo by Martin C. Germano.

A color photograph of this area records a diversity in the objects' structures and colors. NGC 2024 is now a yellow nebula bisected by a curving dust band. NGC 2023, southeast of Zeta, is a blue-white spot of dust reflecting Zeta's light toward us. IC 434, the long, thin strip of red gas, frames the Horsehead Nebula.

Photography shows detail in the Pleiades star cluster, a bright object familiar to every winter stargazer. The cluster is a dipper-shaped group of blue-white beacons. A photograph shows that faint, bluish nebulosity surrounds the cluster's brightest stars, as if they were streaked with a paint brush. The nebulosity is residual dust from the formation of the cluster's stars and is difficult at best to detect with the largest amateur telescopes.

Northeast of the Pleiades is another challenging object for visual observers. The California Nebula in Perseus is a tube of gas heated so it glows by a nearby hot star. The photo above shows the nebula's faint outer extensions, its curving body variously illuminated across its surface, and an extensive network of dark lanes etched into the background stars.

Near the California Nebula in central Auriga lies a complex of nebulosity. The pair of photos on page 98 shows central Auriga in both blue and red light. The blue-light photograph shows star clusters M36 and M38 and resembles a backyard telescope view. The photograph made with red-sensitive film reveals a scattered group of faint nebulae. The largest, the Flaming Star Nebula, surrounds the star AE Aurigae, which appears as a lone star in the blue-light photo.

Enjoy observing the faint and challenging deep-sky objects scattered throughout the night sky. But also explore a second type of observing. Pull out your astronomy books and magazines and observe the wonders of the universe as they appear in photographs. You'll get a new perspective on the objects that you just can't get anywhere else. □

Journey to the Center of the Galaxy

Unlock the splendor of bright deep-sky objects near the galactic center with binoculars and small telescopes.

by Phil Harrington

Galaxies are richly filled with stars, gas, and dust, the glowing embers of energy that bathe the otherwise dark universe in light.

Our Galaxy, the Milky Way, contains countless star clouds and is laced by intricate networks of glowing gas in the slow process of coalescing into stars. Considered by most astronomers to be an Sb spiral galaxy (though now some believe it to be a barred spiral), the Milky Way spans some 100,000 light-years and is estimated to contain about 400 billion stars. Unfortunately our view of the real action in the Milky Way is largely obscured by thick pockets of dust. Rather than sitting near the center of the Galaxy, we are situated about 25,000 light-years out. Nevertheless, with telescopes and our minds working as spaceships of the imagination, we can observe many of the brightest objects toward the center of the Galaxy — the huge clouds of gas, dust, and stars lying in the thick of the action of our home star system.

Our hunting ground will be confined to Sagittarius and Scorpius — two constellations of the zodiac that lie due south in the mid-evening hours during the summer months. Sagittarius is the Archer of mythology but to modern observers its brightest stars form an asterism that appears like a giant teapot. Scorpius, on the other hand, has bright stars that outline the general appearance of a scorpion. The actual galactic center is obscured by dust clouds, but the direction toward that point is known. It is a position on the sky 1.5° southwest of the star X Sagittarii, midway between the bright Lagoon Nebula and the star cluster M6.

Few people would argue that the most spectacular object in the region of the galactic center is the **Lagoon Nebula**, also known as M8 (NGC 6523). The Lagoon Nebula is one of the largest, brightest, and most detailed emission nebulae in the sky. Emission nebulae glow under their own power, their ionized gas excited into luminescence from the energy of young, hot stars buried within.

On a clear, moonless night, M8 is easily visible to the unaided eye as a bright spot in the Milky Way northwest of the teapot's "spout." A 4- to 6-inch telescope reveals a glowing cloud bisected by a dark lane, or "lagoon." Eight- to 10-inch instruments begin to disclose some of the fainter portions of the nebula, while larger telescopes show a wealth of detail.

Begin your visit to M8 by using your lowest-power, widest-field eyepiece. This way you will be

M22
Mag. 5.1   Size 24.0'   by John P. Gleason

M7
Mag. 3.3   Size 80'   by Martin C. Germano

The Lagoon Nebula
Size 90' by 40'   by Tony Hallas and Daphne Mount

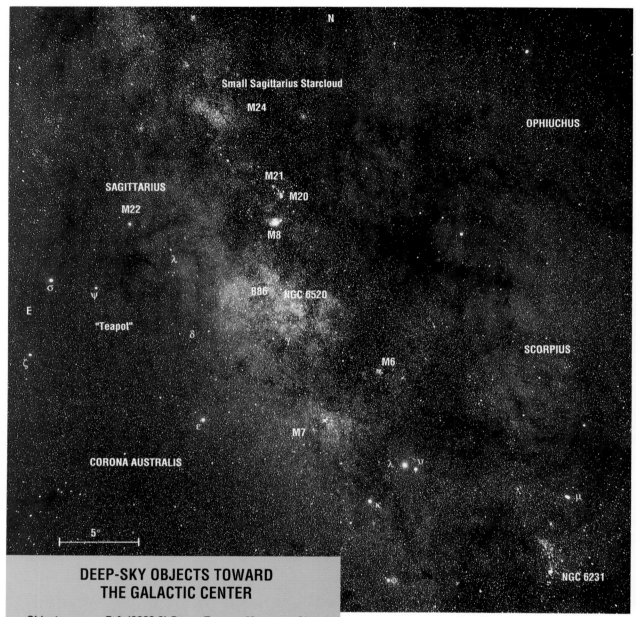

DEEP-SKY OBJECTS TOWARD THE GALACTIC CENTER

Object	R.A. (2000.0) Dec.	Type	Mag.	Size
NGC 6231	16h54.0m −41°48'	OC	2.6	15'
M6 (NGC 6405)	17h40.1m −32°13'	OC	4.2	15'
M7 (NGC 6475)	17h53.9m −34°49'	OC	3.3	80'
M20 (NGC 6514)	18h02.6m −23°02'	EN+RN	—	29' by 27'
B86	18h02.7m −27°50'	DN	—	4'
NGC 6520	18h03.4m −27°54'	OC	7.6p	6'
M8 (NGC 6523)	18h03.8m −24°23'	EN	—	90' by 40'
M21 (NGC 6531)	18h04.6m −22°30'	OC	5.9	13'
B92	18h15.5m −18°11'	DN	—	12' by 6'
B93	18h16.9m −18°04'	DN	—	12' by 2'
NGC 6603	18h18.4m −18°25'	OC	11.1p	5'
M22 (NGC 6656)	18h36.4m −23°54'	GC	5.1	24.0'

Object: Designation in Messier, NGC, or Barnard catalog. **R.A. and Dec.:** Right ascension and declination in equinox 2000.0. **Type:** DN = dark nebula; EN = emission nebula; GC = globular star cluster; OC = open star cluster; RN = reflection nebula. **Mag.:** V magnitude of object. Subscript p = photographic magnitude. **Size:** Diameter in arcminutes.

M24 by George Viscome

Barnard 92
Size 12' by 6'   by Martin C. Germano

able to get most of its full 90' by 40' dimensions inside a single field of view. Then switch to a higher power to zoom in on areas that interest you. The Lagoon is too splendid an object to be fully appreciated with just one look — you'll see more and more in it as one summer season follows another and your eye becomes more practiced. No matter how often I look at M8, I always find something new, something I have never noticed before. One thing I'd like you to notice is the open star cluster **NGC 6530**. It is visible in the eastern part of the Lagoon Nebula. NGC 6530 consists of about two dozen stars ranging from 7th to 9th magnitude and scattered across 15'. Taken by itself, the cluster would be a fine object, but with the added beauty of M8's soft, gossamer clouds, the field is truly striking.

To me, the **The Trifid Nebula**, M20 (NGC 6514) — an emission nebula just 2° northwest of the Lagoon — looks like a celestial flower. The "petals" of the Trifid appear bright through large backyard telescopes (and a narrow-band nebula filter makes them obvious). The Trifid Nebula appears in photos as a vibrantly colored cloud trisected by dark lanes. Six-inch and larger telescopes deliver a view of M20 that approaches these photographs in detail. Through my 8-inch f/7 Newtonian reflector, the nebulosity is bright and obvious. The dark rifts first seen by 18th-century observer William Herschel are visible, first by averted vision only, then directly. Regardless of your telescope, you'll gain the best views of M20 through eyepieces yielding 100x to 150x. While you're in the area, you can take a look at a pretty, bright star cluster. Less than 1° northeast of M20 is the sparkling open cluster **M21** (NGC 6531). Binoculars and finderscopes reveal a blur of stardust speckled with a few 8th-magnitude points of light, while 6-inch telescopes display about half the cluster's 70 stars spread over 13'. Given its distance of 4,000 light-years, the cluster's true diameter must be close to 15 light-years.

After you've enjoyed looking at the Lagoon-Trifid region, set your optics aside and gaze above the teapot's spout with your eyes alone. There, along the stream of the Milky Way, you'll see two clouds of milky light. The largest (and northernmost) is the Scutum star cloud, while the other is the Small Sagittarius star cloud, also known as **M24**. This object is made up of a rich area of thousands of stars appearing to be concentrated together due to our line of sight. Therefore, M24 is best visible in low-power binoculars. From dark-sky observing sites, the splendor of this area is unparalleled in the region: a thick blanket of minute stars lies scattered throughout the eyepiece, crisscrossed by fine, dark nebulae and knots and clumps of stars. Slowly scan the cloud using a short-focus, richfield telescope (RFT), and you'll feel as though you are floating freely among the stars.

Buried within M24 is the tiny open cluster **NGC 6603**. NGC 6603 appears as a dense group of 100 stars, none glowing brighter than 12th magnitude. Though visible in smaller telescopes, NGC 6603 demands an 8-inch instrument to be resolved into stars. After familiarizing yourself with the cloud, look closely at M24 on a dark night and you'll see two dark blotches silhouetted against the glittering backdrop of

The Trifid Nebula
Size 29' by 27' by Tony Hallas and Daphne Mount

stars. This is a pair of dark nebulae — giant clouds of interstellar dust. **Barnard 92** and **Barnard 93** are visible through most telescopes as small, oval "holes" in space. Most observers spot B92 at first glance; the more subtle B93 requires some patience. Like just about all other dark nebulae, B93 is easily visible only from dark observing sites far from all sources of light pollution.

Thus far we've explored the Lagoon-Trifid region and the nearby Small Sagittarius star cloud. Now let's move south along the Milky Way to a point 4° south of the Lagoon to come upon another of Sagittarius' brightest deep-sky splendors. **NGC 6520** is a pretty open cluster set in a striking field that lies some 2° north of Gamma Sagittarii. The group consists of 60 stars, of which 20 to 25 are detectable through an 8-inch telescope. Buried in its center is a bright orange star that shines in sharp contrast against a sea of white and blue-white suns. The NGC field is even more striking because it contains the dark nebula **Barnard 86**, just northwest of NGC 6520. Remember, when you try to observe dark nebulae, it's what you *don't* see that counts. In the case of B86, you don't see any stars within a triangular wedge measuring about 4' across. Its contrast against the brilliant Milky Way star fields forces the black void to stand out even through suburban skies. Try medium power for the best view.

Now that we've surveyed the major sights of southern Sagittarius, let's move over into Scorpius. Toward the stinger of the Scorpion lies a pair of large, bright open clusters that are among the finest objects for binoculars found anywhere in the sky because of their great sizes and brightnesses. **M7** (NGC 6475) is the brighter and larger of the two. The unaided eye sees it as a misty glow about 4° northeast of Lambda Scorpii, the star marking the Scorpion's stinger. Even field glasses show M7 as a beautiful array of stars spanning 80'. Nearly a third of the 80 stars identified as cluster members are brighter than 9th magnitude and therefore visible in 7x binoculars. The brightest star of M7, a yellow type-G sun of 6th magnitude, gleams close to the group's center. Take a moment to savor its subtleties: many fainter cluster stars shine with a yellowish cast, while others appear blue-white.

NGC 6520
Mag. 7.6p Size 6' by Martin C. Germano

M6
Mag. 4.2 Size 15' by Martin C. Germano

NGC 6231
Mag. 2.6 Size 15' by Mike Mayerchek

The naked eye is all you need to detect **M6** (NGC 6405), the other bright cluster in Scorpius, as a smudge of light northwest of M7. Through binoculars, M6 appears as a dazzling open cluster of stardust surrounded by a glittering field. The first thing you may conclude about M6 is that it has an unusual boxy appearance. But when you let your imagination roam, you will see why M6 has been nicknamed the Butterfly Cluster. Do you see two stellar "wings" spreading out from the cluster's more densely packed body? The brightest of the 80 stars in M6 is an orange stellar inferno found east of the cluster's center, (in the left wing). This is **BM Scorpii**, a semi-regular variable star whose brightness fluctuates between magnitudes 6.8 and 8.7. An average of 850 days elapses between two maxima.

Now that we've surveyed the best open clusters of the region, we can move on to one of the largest, brightest, and most observed globular clusters of all — **M22** (NGC 6656). Under ideal conditions, M22 is visible to the naked eye as a hazy spot northeast of Lambda Sagittarii, the top of the teapot's lid. A 3-inch telescope resolves the cluster's brightest stars, while an 8-incher shows an uncountable number of points spanning the face of the cluster. For observers living in the Northern Hemisphere, M22 is rivalled only by the magnificent M13 in Hercules. M22 spans 140 light-years, and may hold as many as a million swarming stars. The cluster is about 10,000 light-years from Earth.

Richly glowing gaseous nebulae, scads of stars clumped into open clusters, and globe-shaped balls of thousands of stars are but a sampling of the wonders awaiting backyard observers as they look toward the center of the Galaxy. On the next clear night, take this article out with your telescope or binoculars and journey far from home toward a place bustling with stellar activity. You'll have a first-class seat on a flight of the imagination to the most thickly populated region of our Galaxy — and you'll capture "live" views of some of the Galaxy's most complex denizens.

These objects are best visible in the summer evening sky.

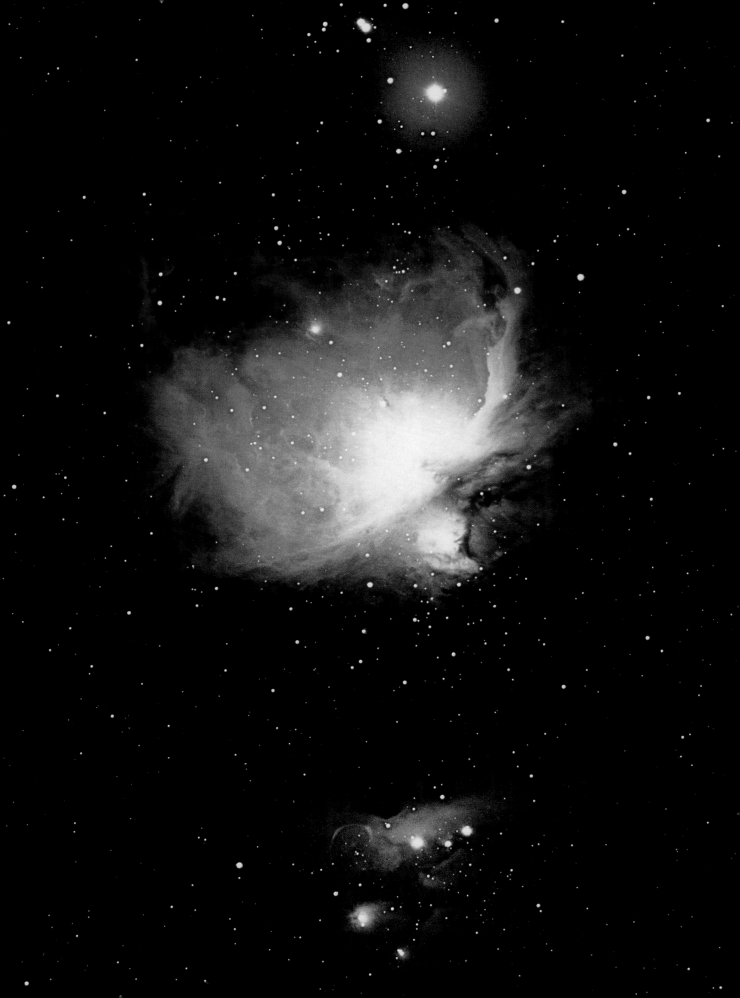

MAGNIFICENT ORION

Filled with bright and beautiful nebulae,
Orion is a constellation not to be missed — especially its gem, M42
by Alan Goldstein

Orion the Hunter is everybody's favorite winter constellation. Orion's star-studded figure contains more 1st-magnitude stars than any other constellation. Its distinctive form, like that of the human frame, is instantly recognizable. Straddling the celestial equator, Orion is well placed for observers throughout both hemispheres.

The most famous inhabitant of Orion is visible to the naked eye as a 4th-magnitude fuzzy patch in Orion's sword. The Orion Nebula (M42, NGC 1976) is an enormous complex of bright nebulosity and dust that is the closest star-making machine to our solar system. Its distance is not precisely known, but the nebula is thought to be between 1,000 and 1,900 light-years away. Consequently, its diameter is between 30 and 45 light-years. The collected mass of M42 is equal to approximately 10,000 Suns.

The Orion Nebula represents the pinnacle of deep-sky observing for amateur astronomers observing in mid-northern latitudes. It is unquestionably the most spectacular emission nebula visible to Northern Hemisphere observers. The nebula spans 85' by 60' and is visible in telescopes as a bright core of nebulosity surrounded by a hauntingly beautiful envelope of faint greenish light arranged in a double-fan shape that fills an entire low-power field. As large and delicate as it appears telescopically, long exposure photography reveals the M42/43 complex is the bright core of an immense expanse of celestial gas that is faintly painted across the region of Orion.

The Orion Nebula has fascinated telescopic observers for nearly four centuries. The first telescopic observation of M42 was made by Nicholas Peiresc in 1611, more than 100 years before Charles Messier included it in his famous catalog of nonstellar objects. Oddly enough, M43, a comma-shaped nebula separated from the larger M42 by a rift of dark nebulosity, was not observed until 1731. The first attempt to sketch the complex beauty of the Great Nebula was that of Christiaan Huygens in 1656. Huygens subsequently called attention to the star system Theta Orionis, now called the Trapezium, in the heart of the nebula. William and John Herschel sought inspiration for their mammoth

Orion by Gregory Field

**M42
Size 85' by 60'   by Mike Sisk**

surveys of the sky by studying complexities of the Great Nebula. In 1880 Henry Draper ensured a key role for the Orion Nebula in the transformation of professional astronomy when it was the first deep-sky object captured on film.

Today, with the advent of large aperture telescopes, M42 has not been lost in the deep-sky shuffle. It is still turned to in awe by more observers than any other deep-sky object. It is a favorite for observers with binoculars, and is spectacular with any telescope. M42 offers something for every instrument: Wide-field, low-power instruments can trace the nebula's faint outer extensions. High-power or long focal-length telescopes can study the intricate detail and texture of M42's innermost regions. Whether you are observing at 7x or 350x, there is always something fascinating to see in M42.

M42 offers numerous observational challenges.

How much nebulosity can you see? Can you detect color? What is the faintest star observable within M42? How many of the variable stars can you see? Are any double stars visible? What you see with your telescope will be determined by many factors.

As an exercise in determining what you see, consider making a sketch of M42. Don't worry about how artistic it looks — you only have to show it to yourself. Over several weeks, sketch it several times. Your sketching ability may not improve, but your ability to see more detail will. Eventually, given a telescope of sufficient aperture, the incredible detail in the gas cloud will exceed your mind's ability to soak up and process the flood of photons you will see. The result is an observation you will never forget. Yes, the nebula is impossible to draw and is significantly more dazzling than the best photograph ever made.

The powerhouse of the Great Nebula is a cluster of very young (approximately 30,000 years old) stars in the center of the cloud. The lower energy photons of light we see are given off as a result of the high-energy ultraviolet radiation given off by the stars and illuminating the gas by fluorescence.

What color is M42? Why doesn't it appear red as it does in photographs? The answer lies within our own eyes. The human eye is not sufficiently sensitive to pick up the red glow emitted by the hydrogen gas in nebulae like M42. While the nebulosity is mostly hydrogen gas, also present are helium, carbon, oxygen, nitrogen, sulfur, and neon. Our eyes pick up the green glow of doubly ionized oxygen — it isn't the most abundant element in the nebulosity, but it emits photons with sufficient energy for our eyes to detect. Consequently, observers capable of detecting color in the Orion Nebula are most likely to see a greenish glow. The intensity and hue varies with aperture and eye sensitivity. Some observers report the Great Nebula as pale green or greenish gray in color. Others call it pea green. With a 20-inch f/5 telescope, it is best described as the green color of fresh spring leaves. A few experienced observers see a light reddish hue in the outer streaks.

John Herschel, the first observer to do an in-depth study of M42, named several features in the nebula. Some of these regions may seem familiar, even if the names are odd. For instance, the bright central area surrounding the Trapezium was named *Regio Huygeniana* after Christiaan Huygens, whom Herschel mistakenly thought discovered the Orion Nebula. The dark nebular protrusion into the area near the Trapezium was called *Sinus Magnus*, the Great Gulf. In his 1844 *Bedford Catalogue*, William Henry Smyth called this area "the fish's mouth." The so-called arms of the nebula that trail to the south and east were broken into two tendrils, *Proboscis Major* and *Proboscis Minor*, the Greater and Lesser Trunks. The darkness between it was dubbed *Regio Messieriana* after the French comet hunter Charles Messier.

M43 is the comma-shaped cloud of gas that is visually separated from M42 by a wedge of dark nebulosity. An 8th-magnitude star in the center causes this part of the nebular complex to glow. Astrophotos show that M43 is part of the larger M42 cloud and that the separation is merely an illusion produced by line of sight.

The **Trapezium**, Theta^1 Orionis, is one of the best known multiple star systems. It is easily resolvable with small instruments. The four stars are designated A, B, C, and D. Star C is the brightest at 5.4 magnitude, while star D is second at 6.3. The other two are variable stars. Star A is an eclipsing binary. Normally at magnitude 6.8, it fades one magnitude every 65.4 days. Star B is an eclipsing binary normally glowing at 7.9 magnitude, with a period of 6.5 days. The four members of this multiple star system are hot and very young: their spectral types are either type O or B. These stars cause much of the Great Nebula to glow.

There are other stars around the Trapezium that offer a challenge to observers. About 4" north of star A is an 11th-magnitude star designated E. It was discovered by F. G. W. Struve in 1826. Some 4" southeast of C lies star F, an 11th-magnitude star discovered by John Herschel in 1830. Star G was discovered by Alvan G. Clark in 1888. It is located about 6" west of star D. At 16th magnitude, the G component is a challenge for larger instruments, made especially difficult by the bright glow of the nebula. In 1888 E. E. Barnard discovered the 16th-magnitude double star designated H, whose components are separated by 1.3". It lies west of C and south of A, forming the vertex of a right triangle with those two Trapezium members.

Theta^2 Orionis lies a few arcminutes southeast of the Trapezium, just outside of the central nebular glow (the Huygenian region). Theta^2 is a wide double star with a separation of 52.5". The brighter star glows at 5.4 magnitude, the companion at 6.8. This easy pair is usually ignored by observers reveling in the beauty of the Great Nebula.

Less than 1° north of the Orion Nebula are two other deep-sky wonders. **NGC 1973-5-7** is a single nebula, each NGC number representing a bright area excited by a bright star. In photographs, the nebula is bordered to the south by an irregular lumpy dust cloud. This nebula is difficult to see under most observing conditions but is visible with an 8-inch telescope under very dark skies. The most difficult patch, NGC 1977,

M42
8-inch f/5 reflector at 55x Sketch by Phil Harrington

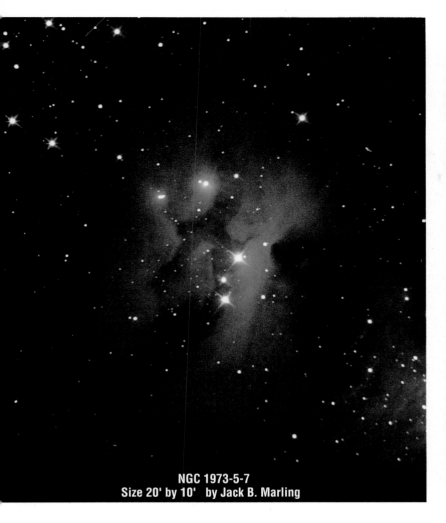

NGC 1973-5-7
Size 20' by 10' by Jack B. Marling

M78
Size 8' by 6' by Jack B. Marling

involves the 4th-magnitude star 42 Orionis. The star's light floods the field, making the nebula much more difficult to observe than the other two parts. The Y-shaped rift in the center of the nebula is very difficult to see. Color photographs reveal that this nebula is a blue color, indicating that it is reflecting much of the starlight.

About 30' north is the coarse open cluster **NGC 1981**. This cluster of less than fifty stars lies within an area about 45' across. This group is best seen with low power. Webb called this cluster "a grand neighborhood" and a "brilliant field containing Struve 750 and Struve 743." These double stars glow at magnitudes 6 and 8 and 6.9 and 8 respectively and have separations of 4.3" and 1.8". The cluster is somewhat elongated, consisting of two rows of stars with a sprinkling of others in the field. NGC 1981 is best observed in 3-inch to 6-inch scopes at relatively low power.

Several deep-sky objects lie south of the Orion Nebula. **NGC 1980** is an emission nebula that is part of the complex in Orion's sword. Like NGC 1973-5-7, NGC 1980 is a pale, ghostly cloud barely visible with most telescopes. It is easy to locate, as it surrounds the bright star Iota Orionis. You will need clean, well-collimated optics 8 inches or more in diameter to pick up this faint wisp of nebulosity. Use averted vision by scanning the field and let the sparse light from NGC 1980 hit the light-sensitive rods in your eyes. Your success in finding this object will be much more likely if you use averted vision.

Our last featured object is **M78** (NGC 2068), a bright, pretty reflection nebula. M78 consists of two nebular patches, the brightest glowing at 10th magnitude. The two patches are separated by a distinct dark lane. With a small telescope, you'll need a dark, moonless night to detect this dark lane.

Long-exposure photos reveal a substantially more complicated scene than is visible in a telescope. Webb called M78 a "singular, wispy nebula." Smyth recorded the nebula as "a singular mass of matter trending from a well-defined northern disk into the south-following quadrant, where it melts away." Many observers think M78 resembles a comet with a double nucleus and a short tail fanning away. The double star in the heart of M78 glows at about 8.2 magnitude.

Don't miss Orion this winter. Not only is it a bewitching sight to the naked eye on frosty nights, but it also offers a beautiful array of deep-sky objects for your telescope. Delve deep into studying the intricacies of M42, one of the most fantastic objects in the sky. But don't neglect Orion's other deep-sky wonders, hot spots in an intricate network of nebulosity that covers the entire constellation.

These objects are best visible in the winter evening sky.

The Wonders of the Coma Cluster

Coma Berenices contains a star cluster you can see with the eye alone and also a rich group of galaxies to savor with your telescope.

by Alan Goldstein

This spring a great opportunity to explore deep-sky objects awaits you. With the naked eye and binoculars, you can view the star cluster making up the constellation Coma Berenices. With a telescope, you can peer far deeper into the realm of its galaxies. You can even explore the faint light of a remote cluster of galaxies. All you need is a free evening, the star map in this article, and a little patience.

Coma Berenices is unique among its peers. As constellations go it is small, and it's not even bright enough to be visible from moderately light-polluted areas. Yet its abundance of galaxies won't be found in many larger constellations. Most of its bright stars are members of the same star cluster. The star cluster isn't as compact as the Hyades or Pleiades because it is relatively nearby.

As you fix your gaze between Leo and Bootes, the first thing you'll see is a fuzzy patch of light several degrees across. This is the combined light from 50 or more stars in the **Coma Berenices star cluster** (Melotte 111). Just 250 light-years away, the Coma star cluster is in our cosmic backyard. In fact it is the third closest star cluster to Earth (after the Hyades in Taurus and the Ursa Major moving group, which is so close it does not look like a cluster at all). As we see the Coma star cluster, its members are strewn over an area of sky measuring 6° across, or twelve diameters of the Full Moon. Because the cluster is so large, it is best observed with a good pair of 7x50 binoculars.

Several members of the Coma star cluster are double or multiple stars. For example, **12 Comae Berenices** is a pretty triple star with a combined magnitude of 4.7. It is easy to find, being the brightest star on the western side of the cluster. Its companions are located 12" and 35" from the primary and glow at magnitudes 8.5 and 11.5, respectively. Another member of the cluster is **17 Comae Berenices**, a wide pair separated by 145". The brighter member shines at magnitude 5.4, the secondary at 6.7. The magnitude 6.7 star has a close companion separated by less than 2". At 13.7 magnitude, this faint star requires an 8-inch to 10-inch telescope to be seen. The star **24 Comae Berenices** is a beautiful multiple star system composed of magnitude 5.0 and 6.5 stars. To find this star, start at 17 Comae Berenices and star-hop 4° east-southeast to 23 Comae Berenices, then 4° due south to 24 Comae Berenices. Separated by 20.3", the yellow and blue stars in this pair are reminiscent of Albireo in Cygnus. Although

NGC 4725   Mag. 8.9   Size 7.5' by 4.8'
by Rick Schrantz

slightly fainter that Albireo, 24 Comae Berenices is gorgeous in small telescopes.

Coma Berenices is best known for the many galaxies that lie within its borders. The finest is unquestionably the edge-on spiral **NGC 4565**, located less than 2° east of 17 Comae Berenices. At magnitude 10.2, this Sb-type galaxy is visible in small telescopes or a good pair of binoculars under excellent skies. Its dimensions are 15' by 1.1'. When you observe NGC 4565, you're now looking far beyond the 250-light-year distance of the Coma Star Cluster. NGC 4565 lies a whopping 30 million light-years away.

When you observe NGC 4565 you'll notice the galaxy is rich in dust. Many spiral galaxies contain large quantities of dark interstellar dust, and the most prominent feature of NGC 4565 is its thin central dust lane, which bisects this spindle-like object. Careful scrutiny reveals the galaxy's nucleus just north of this dust lane. In a 2-inch telescope this galaxy is visible as a slender ribbon of light. In a 4-incher you'll see the dark lane distinctly. In a 6-inch instrument you'll see the asymmetric nature of the dust lane and the galaxy's

NGC 4565   Mag. 10.2   Size 15' by 1.1'
by Tom Eby

small, bright nucleus. A large telescope reveals subtle details in the dark lane.

Although they are not as spectacular as NGC 4565, many other galaxies lie nearby. You'll find **NGC 4494** just 1° west of NGC 4565. If you're using a small telescope, NGC 4494 might be passed by as a slightly out-of-focus star. It is located in the same field of view as 17 Comae Berenices. Like most elliptical galaxies, NGC 4494 bears magnification well — don't be afraid to try 150x or 200x to examine its details. **NGC 4448** lies just 3.5° northwest of NGC 4565 and within the same low-power field of view as the bright star Gamma Comae Berenices. NGC 4448 glows at magnitude 11.4 and measures 2.9' by 1.0' across. This compact type-Sb spiral galaxy is visible as a pale oval orb in a 3-inch telescope under superb skies and in a 4- or 5-inch scope under less than ideal conditions.

Also near NGC 4565 lies one of the largest galaxies in the constellation, **NGC 4559**. To find this galaxy simply center your scope on NGC 4565 and move 2.5° north in declination, to a point 2° east of Gamma Comae Berenices. NGC 4559 is a hefty object measur-

NGC 4559   Mag. 9.9   Size 11' by 4.5'
by Martin C. Germano

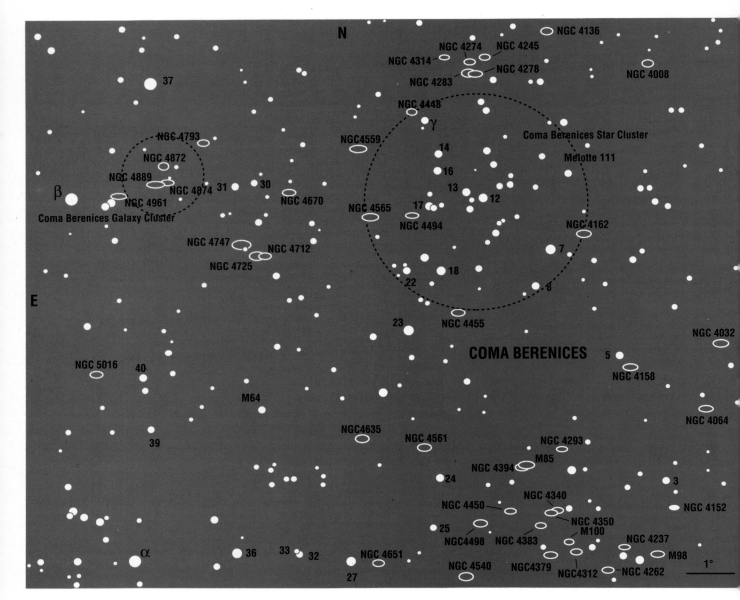

NGC 4725 and NGC 4712   Mags. 8.9 and 13.5
Sizes 7.5' by 4.8' and 1.9' by 0.8'
by Tony Hallas and Daphne Mount

ing 11' by 4.5'. NGC 4559 is a highly inclined Sc-type spiral galaxy with a very weak brightening toward the center. Otherwise, it is evenly illuminated across its surface. Three foreground stars are superimposed on this galaxy.

Located about 2° south of 31 Comae Berenices is **NGC 4725**. This magnitude 8.9 spiral has a weak bar-shaped hub, which in the galaxy classification scheme translates to a type S(B)b. This is a transition between barred spirals with a distinct bar-shaped hub and spirals with arms forming at various intervals around the disk. The photos of transitional barred spirals typically show arms emerging at various points along a bar-shaped central hub. A small telescope reveals a bright nuclear region in NGC 4725 surrounded by a mottled glow spanning 7.5' by 4.8'. A large telescope reveals more detail, including the galaxy's elongated hub and faint spiral structure.

NGC 4712 is located less than 15' west of NGC 4725, and both galaxies are visible in the same field of view. At magnitude 13.5, this compact, oval-shaped spiral galaxy requires a 10-inch telescope to be seen.

Coma Galaxy Cluster
by Joe Liddell

Coma Galaxy Cluster
by Tony Hallas and Daphne Mount

Deep-Sky Objects in Coma Berenices

Object	R.A. (2000.0)	Dec.	Mag.	Size/Sep.
Melotte 111	12h25m	+26°	1.8	275'
12 Com	12h22.5m	+25°51'	4.7, 8.5	12'
NGC 4448	12h28.2m	+28°37'	11.4	2.9' by 1.0'
17 Com	12h28.9m	+25°55'	5.4, 6.7	145'
NGC 4494	12h31.4m	+25°47'	9.9	4.8' by 3.8'
24 Com	12h35.1m	+18°23'	5.0, 6.5	20.3'
NGC 4559	12h36.0m	+27°58'	9.9	11' by 4.5'
NGC 4565	12h36.3m	+25°59'	10.2	15' by 1.1'
NGC 4712	12h49.6m	+25°28'	13.5	1.9' by 0.8'
NGC 4725	12h50.4m	+25°30'	8.9	7.5' by 4.8'
NGC 4747	12h51.8m	+25°47'	12.4	3.6' by 1.4'
NGC 4874	12h59.6m	+27°58'	11.9	2.7' by 2.7'
NGC 4889	13h00.1m	+27°59'	11.4	3.0' by 2.1'

Object: Designation in NGC or other catalog. **R.A. and Dec.:** Coordinates in equinox 2000.0. **Mag.:** V magnitude. **Size/Sep.:** Size in arcminutes or separation in arcseconds.

Smaller scopes (6 to 8 inches) may reveal NGC 4712, but only under very good skies. This 1.9' by 0.8' galaxy is a good test for moderate apertures. Also in the same field, **NGC 4747** can be found about 20' northeast of NGC 4725. This highly elongated galaxy is number 159 in Halton C. Arp's *Catalogue of Peculiar Galaxies*. Photographs reveal an edge-on, dusty galaxy that is presumably a spiral. There are several very faint plumes of galactic material splashed away from the disk. This may be the result of an interaction with another galaxy in the area. A large telescope reveals some mottled patterns of dust along the narrow disk of NGC 4747. The plumes themselves are beyond the reach of amateur instruments.

If you have a small telescope you'll find the next object challenging. The **Coma Berenices Galaxy Cluster** is a swarm of more than thirty faint galaxies visible in small telescopes. At 11th magnitude, **NGC 4889** is the cluster's brightest individual galaxy. As you peer into the eyepiece, consider that this 11th-magnitude smudge is the combined light of hundreds of billions of stars located over 425 million light-years away — some 20 times more distant than the Coma-Virgo cluster of galaxies. This object deserves at least a glance. Chances are, if your telescope can pick up NGC 4889, it will allow you to glimpse nearby **NGC 4874**, which is a half magnitude fainter.

NGC 4889 and 4874 are both giant elliptical galaxies surrounded by a swarm of smaller elliptical and lenticular galaxies. An 8-inch instrument reveals that NGC 4889 is elongated in an east-west direction. A 10-incher shows an oval glow with a brighter center. Larger telescopes allow you to see some (or all) of the thirteen other galaxies in the immediate vicinity of NGC 4889. As impressive as that sounds, you can see more if you slide on to NGC 4874. There, you will see a swarm of seventeen dimmer galaxies. Both fields are a must see for observers with 10-inch or larger scopes. Overlook this field of wonder and you are truly missing a grand view of the universe!

Whether you scan the objects in Coma Berenices from mere tens of light-years or hundreds of millions of light-years, each photon your eyes catch is a moment in our own Earth's history as well as the history of the universe. Starlight from members of the Coma star cluster began its journey before the American Revolution. The light of the Coma galaxy cluster dates back to the time in Earth's history when primitive plants and animals were the pinnacle of evolution. Imagine — all within the boundaries of a tiny, mild-mannered group of stars! □

These objects are best visible in the spring evening sky.

EXPLORE THE SOUTHERN SKY

Rich with nebulae, clusters, and the dazzling Magellanic Clouds, this region contains some of the most spectacular objects in the heavens.
by Gregg D. Thompson

Where can you find the finest globulars, the brightest and most detailed galaxies, the richest open clusters, the brightest and most impressive nebulae, and the largest and darkest dark nebulae? In the southern sky.

From about 30° south latitude the center of our Galaxy passes directly overhead, providing exquisite views of all types of galactic objects. The rich Centaurus arm of the Galaxy lies in the southern sky and offers spectacular views of open clusters, the Eta Carinae Nebula, and the dusty dark nebula called the Coal Sack. The southern sky also holds the biggest and brightest globular clusters and the Magellanic Clouds, our galaxy's satellites. Whether you visit the Southern Hemisphere on a vacation or observe the southern sky year round, there's an awful lot to see.

The best object in the southern sky is the **Large Magellanic Cloud,** the largest satellite galaxy to the Milky Way. Near the center of the LMC lies the **Tarantula Nebula** (NGC 2070), the brightest object in the LMC. This object is completely unlike anything visible in the northern sky: A 12.5-inch scope casually aimed near its center shows a bright, complex structure with tentacles of nebulosity that are both bright and faint and narrow and broad.

As the field drifts, you see scores of other gaseous nebulae, open star clusters, and globulars. One region that can be glimpsed with a keen, unaided eye is the group of nebulae designated **NGC 1760-9**, a large cloud containing several bright star clusters in three basic clumps. Another fascinating nebula is **NGC 2032-5,** the Seagull Nebula. This object is shaped like a seagull about to dive.

Unfortunately, the Milky Way's other satellite galaxy, the **Small Magellanic Cloud,** is often overlooked, although it too contains unusual clusters and nebulae. The SMC's answer to the Tarantula Nebula is the bright nebula **NGC 346**. NGC 346 can be glimpsed with the naked eye under perfect conditions. In my 12.5-inch scope at high power the nebula fills over half of the field of view. Nearby lies **NGC 371**, a bright star cluster best viewed with a nebular filter, which shows an almost perfectly circular nebulosity enveloping the cluster.

The beautiful globular cluster **NGC 362** lies near the tail of the Small Magellanic Cloud. NGC 362 appears so rich in a 12-inch or larger scope that it confounds those who want to sketch it. Thousands of glowing points crowd inside a high-power field of view. Although NGC 362 is smaller and dimmer than Omega Centauri, its center is far more concentrated. Nearby **47 Tucanae** (NGC 104), a globular belonging to the Milky Way, is one of the two finest globular clusters in the sky. The outer stars in this cluster appear much fainter than the ones in close to the center. The core of 47 Tucanae is so jammed with stars that it appears like one globular laid directly upon another. The cluster's core stars are strongly yellow in color.

The most beautiful cluster in the southern sky is the elliptically shaped **Omega Centauri** (NGC 5139). This object is easily visible with the naked eye as a fuzzy "star." In binoculars it is comet-like and was often confused with Comet Halley when the comet passed near it in April 1986. Backyard telescopes resolve thousands of Omega's estimated one million stars. Near its broad central region are three small, dark oval patches silhouetted against the stardust background. The two most prominent form owl-like "eyes."

**The Southern Milky Way in Crux
by Chris Floyd**

The Eta Carinae Nebula
Size 120' by 120'   by Ronald Royer

Peculiar Galaxy NGC 5128
Mag. 7.0   Size 18.2' by 14.5'   by Jack Marling

This feature has no physical significance, however; it is formed by chance arrangements of bright stars.

Several other southern globulars are worth a look. The impressive globular **NGC 6397** lies in the constellation Ara. Although not the largest globular in this constellation, NGC 6397 is rich, bright, and condensed. **NGC 6752** in Pavo lies beside an 8th-magnitude star. This cluster can be resolved into many members in a small telescope, but its prominent background glow is evidence that countless faint stars remain unresolved. **NGC 4833** in Musca has two distinctive areas without bright stars, making it appear that a "V"-shaped incision has been cut into the globular's core. **NGC 5286** lies beside the naked-eye star M Centauri. **NGC 2808** in Carina is rich and compressed to a brilliant core that looks like a more distant version of 47 Tucanae.

One of the most peculiar objects in the southern sky is the spherical galaxy **NGC 5128** (Centaurus A). This object appears like a large ball of light bisected by a prominent dust band. In telescopes larger than 12.5 inches in aperture, the double nature of the dust lane can easily be seen. Radio telescopes pointed at Centaurus A "hear" a rumble that astronomers think originates from a black hole at its core.

Another beautiful sight that lies nearby is the edge-on galaxy **NGC 4945.** This object is faintly visible in binoculars; in an 8-inch telescope it extends from one side of the field to the other, spanning nearly 30'. Pavo's **NGC 6744** is a huge face-on spiral. This galaxy's arms are too faint to be easily recognizable in small telescopes, but an 18-inch or 20-inch scope does the trick.

The Jewel Box Cluster
Mag. 4.2   Size 10'   by Gregg D. Thompson

Omega Centauri
Mag. 3.7   Size 36.3'   by Jack Newton

The Tarantula Nebula
Size 40' by 25'   by Chris Floyd

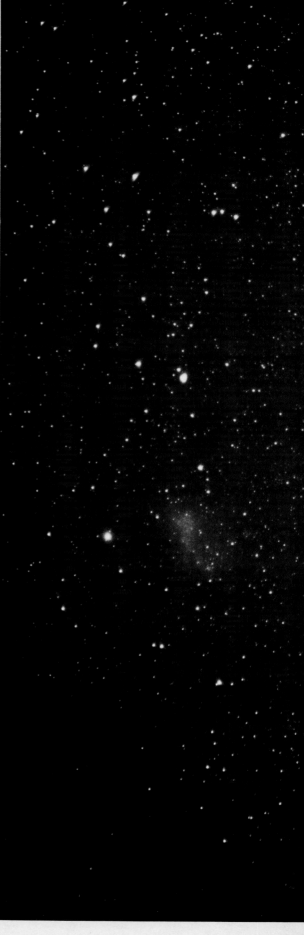

Perhaps the most spectacular galactic object in the southern sky is the **Eta Carinae Nebula** (NGC 3372), a giant rose-colored emission nebula that makes M42 in Orion look like kid's stuff. Prominently visible with the naked eye, the Eta Carinae Nebula fills a finder scope with glowing nebulosity extending over 2° in diameter. Placing an O III filter in my 12.5-inch scope helps me see the Eta Carinae Nebula as a large, bright complex of nebulosity involved with bright stars. Strands and wisps of light spanning a great range of surface brightness — from intensely bright to nearly imperceptible — dance across numerous fields of view.

At the center of the Eta Carinae Nebula lies the **Keyhole Nebula**, a small, very opaque dark nebula surrounded by extremely bright nebulosity. Surrounding the brightest star in the Eta Carinae Nebula — Eta Carinae itself — is a curious feature that seems to be a blur caused by poor optics to many first-time observers. This is in fact the **Homunculus Nebula**, a ruddy glow shaped like a bow-tie. This object is bright enough that it is visible as a rusty-red nebula even in an 8-inch scope. The Homunculus Nebula is best observed with high magnification under good seeing.

Many remarkable star clusters lie in the southern sky. There are many extremely fine, mostly unresolvable, naked-eye open clusters within 20° of Eta Carinae. **NGC 2516** is a dead ringer for the Beehive Cluster and needs only the slightest optical aid to resolve its one hundred stars. **IC 2391** and **IC 2602** are partially resolvable with the unaided eye and are best seen with small binoculars. **NGC 3114**, only 4° west of Eta Carinae, resolves into hundreds of faint stars in apertures larger than 10 inches. **NGC 3532** displays a bright

The Large and Small Magellanic Clouds
Mags. 0.1 and 2.3   Sizes 650' by 550' and 280' by 160'   by Ronald Royer

The Southern Sky

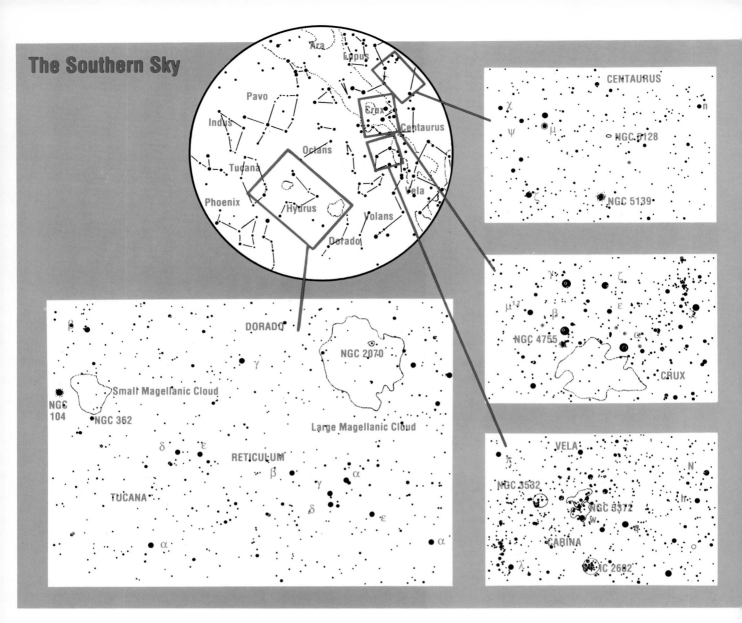

Wonders of the Southern Sky

Object	Type	R.A.(2000.0)	Dec.	Mag.	Size
NGC 104	GC	0h24.1m	−72°05'	4.0	30.9'
SMC	Gal	0h52.7m	−72°50'	2.3	280'x160'
NGC 362	GC	1h03.2m	−70°51'	6.6	12.9'
LMC	Gal	5h23.6m	−69°45'	0.1	650'x550'
NGC 2070	EN	5h38.7m	−69°06'	—	40'x25'
NGC 2516	OC	7h58.3m	−60°52'	3.8	30'
IC 2602	OC	10h43.2m	−64°24'	1.9	50'
NGC 3372	EN	10h43.8m	−59°52'	—	120'x120'
NGC 3532	OC	11h06.4m	−58°40'	3.0	55'
NGC 4755	OC	12h53.6m	−60°20'	4.2	10'
NGC 5128	Gal	13h25.5m	−43°01'	7.0	18.2'x14.5'
NGC 5139	GC	13h26.8m	−47°29'	3.7	36.3'
NGC 6397	GC	17h40.7m	−53°40'	5.7	25.7'

Object: Designation in NGC, IC, or other catalog. **Type:** Object classification; GC - globular cluster, OC - open cluster, EN - emission nebula, Gal - galaxy. **R.A. and Dec.:** Right Ascension and Declination for equinox 2000.0. **Mag.:** V magnitude. **Size:** Dimensions in arcminutes.

orange star in its center and another one at its edge. A dark void divides this cluster and makes it appear like a glowing arrowhead. **NGC 3766** is so compact that it looks like a 5th-magnitude star to the naked eye. It is rich with bright stars with several hues of blue and red. **NGC 3293** competes with **NGC 4755,** the famous Jewel Box cluster. Both are bright, compact star groups with two orange-red stars surrounded by a blue-white smattering of stars.

With so many outstanding examples of clusters, nebulae, and galaxies visible from the Southern Hemisphere, it's certainly worth a visit at some point in your life. As a visitor your observing time will be precious so list the things you want to see and plot them on an atlas ahead of time. But be warned: After seeing what the southern sky holds, you may never want to return to the North!

To see these objects you must observe from a southern latitude.

The Coal Sack and Crux
photo by Ronald E. Royer

Observing the Andromeda Galaxy

The great spiral in Andromeda offers it all to galaxy observers — dust lanes, a concentrated nucleus, star clouds, globular clusters, and a bevy of unusual satellite galaxies.

by Alan Goldstein

It is visible on dark autumn nights as an elongated smear of gray-green light stretching over two diameters of the Moon. Under exceptionally transparent skies it appears almost three-dimensional: a softly glowing almond hanging far above the bright Square of Pegasus. It is the single most distant object visible to the naked eye, the Andromeda Galaxy. This nearby spiral galaxy — a twin of our own Milky Way in the Local Group of galaxies — lies 2.2 million light-years deep into the night sky. Its light, visible on any clear night from late summer to early spring, shows us the nearby spiral as it appeared 2.2 million years, about 1/100 of a galactic year, ago.

Viewing the Galaxy

How much of the **Andromeda Galaxy** (M31, NGC 224) can you see with the naked eye? Under very dark, transparent skies you can trace the extent of the galaxy's arms to about 3° across. (Remember, the diameter of the Full Moon is only one-half degree.) The slightest optical boost such as that given by low-power binoculars will make the oval shape immediately conspicuous. The galaxy appears much brighter in binoculars. Try to determine where M31 ends and the night sky begins — it's nearly impossible. While the light from the galaxy eventually blends into the background light of the sky, the gravitational pull of the Andromeda Galaxy extends across the Local Group. Its presence influences, at least on a small scale, our own Galaxy. M31 more directly holds in its gravitational embrace four satellite galaxies visible in backyard telescopes. We'll examine these shortly.

What does this cosmic neighbor look like when viewed from the backyard? A small telescope shows a bright central hub surrounding M31's nucleus. The galaxy's faintly visible spiral arms form a ghostly glow around the bright hub. A long focal length telescope reveals the glow of the spiral disk against an inky black sky. Those of you with 6-inch to 10-inch scopes will find a glorious view awaiting. You will see streaks of dark, star-obscuring dust, a bright, ball-like nucleus, and a single bright star cloud.

M31's nucleus may be seen as a small fuzzy stellar patch in the center. It bears magnification well because of its high surface brightness. The heart of the Andromeda Galaxy beats with the light of hundreds of thousands of stars packed into a volume no larger than that containing the nearest stars. Inhabitants of this region would live under a "night" sky full of stars, each with the brightness of the Full Moon.

M31, M32, and NGC 205   6-inch reflector at 30x
Sketch by Jeffrey Corder

The Andromeda Galaxy
Mag. 3.5   Size 178' by 63'   by Tony Hallas and Daphne Mount

The Andromeda System

The Andromeda Galaxy (M31, NGC 224, UGC 454)
Sb I-II galaxy
R.A. 0h42.7m, Dec. +41°16' (equinox 2000.0)
Mag.V = 3.5
Size = 178' by 63'

M32 (NGC 221, UGC 452, Arp 168)
E2 galaxy
R.A. 0h42.7m, Dec. +40°52' (2000.0)
Mag.V = 8.2
Size = 7.6' by 5.8'

Photo by Bill Iburg

M32

NCC 206

NCG 205

NGC 205 (UGC 426)
E6: galaxy
R.A. 0h40.4m, Dec. +41°41' (2000.0)
Mag.V = 8.0
Size = 17.4' by 9.8'

Photo by Preston Scott Justis

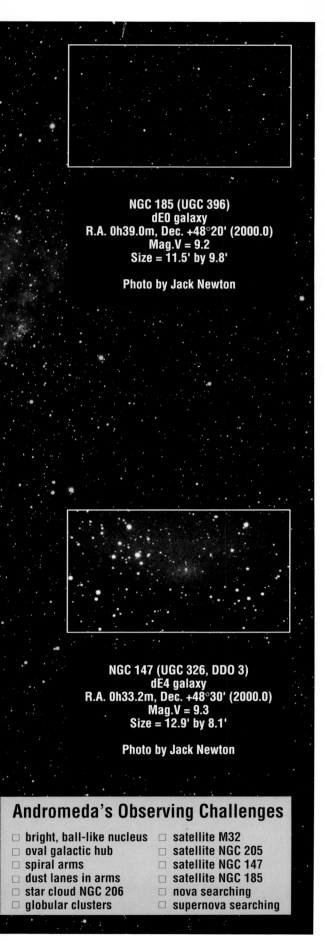

NGC 185 (UGC 396)
dE0 galaxy
R.A. 0h39.0m, Dec. +48°20' (2000.0)
Mag.V = 9.2
Size = 11.5' by 9.8'

Photo by Jack Newton

NGC 147 (UGC 326, DDO 3)
dE4 galaxy
R.A. 0h33.2m, Dec. +48°30' (2000.0)
Mag.V = 9.3
Size = 12.9' by 8.1'

Photo by Jack Newton

Andromeda's Observing Challenges

- ☐ bright, ball-like nucleus
- ☐ oval galactic hub
- ☐ spiral arms
- ☐ dust lanes in arms
- ☐ star cloud NGC 206
- ☐ globular clusters
- ☐ satellite M32
- ☐ satellite NGC 205
- ☐ satellite NGC 147
- ☐ satellite NGC 185
- ☐ nova searching
- ☐ supernova searching

With an 8-inch or larger scope, the most visually interesting feature in M31 is the dark nebulosity surrounding the brightest spiral arms. The foreground portion of the spiral disk will be discernible from the dark lanes. The dust lanes that are easiest to see lie in front of the central hub.

Large Telescope Challenges

Large backyard scopes (12 inches and more aperture) pull in more light, but at the expense of a smaller "window" to look through. Instead of observing the galaxy as a whole, you must look at it piecemeal. While small telescopes can pick out dark nebulae, larger instruments can show brighter condensations of light. Apparently less aperture is necessary to see areas where there are fewer stars than areas where there are more stars! (Actually the dark nebulae are substantially larger than the stellar associations, clusters, and emission nebulae that create the bright portions of the galaxy's spiral arms.)

The Andromeda Galaxy does contain one massive concentration of stars that is comparable in size to some of the large dark nebulae. **NGC 206** is an extremely large cloud of young, hot stars and glowing nebulosity that measures thousands of light-years across. Under excellent skies, NGC 206 can be seen with averted vision in telescopes as small as 6 inches in aperture. It is easier to pick out in larger rich-field telescopes — knowing where to look is half the battle (see the photos at left and on page 80).

Other challenges await large telescope observers. Searching for novae in the central area of M31 might make a nice challenge. Within a 15' radius of the nucleus, Halton C. Arp found as many as 30 novae during a year and a half of photographic patrolling. Such novae may brighten to 15th magnitude, and nightly monitoring may reveal one or two each month. The sheer number of foreground stars will make the task difficult at first, but a good large-scale photo like the one at left will help you spot the transient interlopers.

Other luminous stars lie just beyond the visual range of large telescopes. These stars can be resolved photographically with the large telescopes and hypered fine-grain film available to many of today's amateur astronomers. Individual Cepheid variables, first used by Edwin Hubble to determine the distance of the Andromeda Galaxy, are beyond the range of all but the largest amateur telescopes. Using charts developed by astronomers (such as the *Atlas of the Andromeda Galaxy* by Paul Hodge), amateur astrophotographers will be amazed at the vast amount of detail that can be resolved using state of the art deep-sky photographic techniques.

The challenge of supernova hunting lies in exercising patience more than anything else. The last supernova, given the variable star designation S Andromedae, blazed up to brighter than 6th magnitude in August 1885. It was not observed until a few days after maximum. S Andromedae was located near the nucleus. Because of the high density of stars in that region of the galaxy, other supernovae are likely to be seen there as well. It is unlikely that the next supernova in M31 will be found after it peaks in brightness. The challenge for an observer will be to see and report it first.

The Andromeda Galaxy is full of other interesting

**Globular Clusters in the Andromeda Galaxy
by Paul Roques**

features. For dedicated observers looking for something different, M31 offers a splendid array of globular clusters. Granted, most of these objects require large telescopes to be seen. They are too distant to be seen as anything more than faint stars, the brightest glowing between magnitudes 12 and 13. See the photo above, which will serve as an excellent finder chart for many of the brightest globulars in M31.

The Andromeda Satellites

Astronomers know of eight major satellite galaxies belonging to the Andromeda Galaxy. Half of these are too faint to be picked up with amateur telescopes. The other four are well known and, although low-mass objects, are visible without difficulty in small telescopes.

M32 (NGC 221) and **NGC 205** are dwarf elliptical galaxies. Both are small and contain just a small fraction of the stars of their larger spiral companion. They consist primarily of old stars with a preponderance of stars in the cool end of the spectrum. Physically the two galaxies differ only in shape. There are more similarities than differences. Yet, these two objects do not appear identical in a telescope. Why?

One characteristic of a galaxy that is determined by telescopic observation is the surface brightness. This is the amount of light measured in a given unit of area in the sky. The surface brightness is more difficult for observers to comprehend than the simple magnitudes used to measure starlight. As a measure of a galaxy's brightness, surface brightness is more meaningful than magnitude because it takes into account the galaxy's size and brightness. A galaxy that is 8' by 5' may have the same integrated magnitude as a 10' by 8'; yet comparing the two galaxies, the larger one will appear fainter. This is because the light is spread over a larger area. It is fainter per square arc-minute than the smaller galaxy.

In the Andromeda system, M32 and NGC 205 appear different because NGC 205 has a lower surface brightness. It is possible that the stellar density is lower. With approximately the same number of individual stars spread more thinly over a larger area, the galaxy will appear dimmer. It is for the same reason that the arms of Messier 31 are more difficult to see than the central hub. (There is also more dark nebulosity which obscures much of the starlight in this or any other spiral system.)

The closest companion to M31 is M32, the brightest satellite galaxy. It is a dwarf elliptical (E2), with a high surface brightness. Glowing at 8.0 magnitude, M32 can be seen in large binoculars under good dark skies and appears to be embedded in the glow of the larger spiral galaxy. M32 covers a relatively large area of the sky compared with most galaxies (7.6' by 5.8' in diameter), because it is so close. At the distance of some of the galaxies in the Virgo cluster (some 50 million light-years away), a galaxy of M32's physical size would appear very tiny and extremely faint.

Deep-sky gazers may see M32, but many observers overlook it as they seek the grandeur of the Andromeda Galaxy. This galaxy warrants closer scrutiny. While it does not offer the detail available in its larger neighbor, M32 does show a stellar nucleus without a conspicuous central hub. Under very black skies, with the proper equipment, the boundary between this and the larger spiral galaxy is indeterminate and the starlight from both intermingle. M32 lies suspended in front of the giant wheel of stars that gravitationally holds it close.

NGC 205 is sometimes called M110 because Charles Messier did observe it, although he didn't assign a number to NGC 205. This galaxy's visual magnitude is 8.2, but it is more difficult to see than M32 because of its low surface brightness. Its glow is spread over an area of 17.4' by 9.8', nearly four times that of

M32. The result is a deceptively dim galaxy.

NGC 205 is peculiar in that it has several dark dust patches near its center. Not surprisingly, there are a few relatively young blue stars associated with this dust. This is because much of this dust is being compressed by gravity to form new stars. Most elliptical galaxies formed billions of years ago with very little additional new star development. The dark patches are difficult to see visually.

NGC 205 can be seen with binoculars under dark skies. In small telescopes it appears as an elongated glow with a just a slight brightening toward the center. Larger instruments show a brighter central core, but there is no condensed nucleus like M31 or M32. How well can you see NGC 205 under suburban light-polluted skies?

The 15th-magnitude globular cluster designated "G73" belongs to NGC 205 rather than the Andromeda Galaxy. It should be visible with a 6-inch telescope under excellent skies, though a finder chart will be necessary to distinguish it from the many other faint foreground stars in this area.

If you don't find the Andromeda Galaxy's neighboring companions much of challenge, I am not through. Lying 6° to the north, in Cassiopeia, lie two more galaxies gravitationally bound to the Andromeda system. **NGC 147** and **NGC 185** have surface brightnesses comparable to that of NGC 205, although they are somewhat smaller and at least a full magnitude fainter. Be sure you can see NGC 205 quite easily before attempting to observe either galaxy.

NGC 147 and NGC 185 can be found by star hopping west of Omicron Cassiopeiae. NGC 185 lies 1° west of Omicron and is the brighter of the two at 9.5 magnitude. NGC 185 is a dE0 galaxy that measures 11.5' by 9.8' across, making it visible in a 6-inch telescope under very dark skies. Like NGC 205, this compact galaxy has several small dark patches that contain young blue stars, indicating a minor burst of star formation.

NGC 147 lies a degree west of NGC 185. It glows softly at 9.5 magnitude. With a diameter of 12.9' by 8.1', this dwarf E4 galaxy is considerably more elongated than neighboring NGC 185. NGC 147 has a lower surface brightness and requires very dark skies — moonlit or light-polluted skies will make observing this galaxy difficult. Save this galaxy for really dark skies. Once you become familiar with them, try again under less than ideal conditions and notice how difficult they become!

The Andromeda Galaxy and its companions offer skygazers a choice of observing challenges varying from using the naked eye to the largest telescope money can buy. You can simply sit out in the cool, crisp autumn nights and gaze up at the most distant object visible to the eye, daydreaming about what your own neighborhood would have been like over two million years ago.

Or if you like challenges, the Andromeda system provides a veritable potpourri of choices. You can test your ability to pick up low-surface-brightness objects like NGC 147, NGC 185, and NGC 205. You can look for spiral arm structure and dark lanes in M31. You can search for the tiny globular clusters swarming around M31 like bees around a hive. Or you can sit back, relax and soak up the Andromeda Galaxy's beauty just for the pure pleasure of stargazing. Never have so many observers had such a wonderful set of decisions to make.

The Andromeda Galaxy
Mag. 3.5 Size 178' by 63' by Bruce D. McDonald

These objects are best visible in the autumn evening sky.

Galaxy Hunting around the Big Dipper

Explore the spiral arms and dust lanes
of bright springtime galaxies around the Big Dipper.
by Alan Goldstein

Galaxies are the most varied, beautiful objects visible in backyard telescopes. They range from naked-eye visibility to dim smudges barely visible in big telescopes, from graceful spirals to featureless blurs. Besides stars, galaxies are the most abundant class of objects in the sky — roughly 10,000 are visible in amateur telescopes. Springtime offers an especially good opportunity for galaxy hunting, because the evening sky turns away from the cluttered winter Milky Way and lets us see far out into the depths of the universe.

The finest galaxy by far in the spring sky is **M51** (NGC 5194), the Whirlpool Galaxy in Canes Venatici. Called the Whirlpool because of its distinctive spiral pattern, M51 glows at magnitude 8.4 and spans 11' by 8', which makes it one of the brightest and largest spiral galaxies in the sky. Because M51 is almost exactly face-on to our line of sight, its spiral pattern and other details are easily visible.

As if that weren't enough, M51 is the brightest component of an interacting pair. **NGC 5195**, a small peculiar galaxy, is interacting with the outer part of M51 and distorting both galaxies in the process. Two- to 6-inch telescopes show the galaxies as a double patch of misty light. Eight-inch or larger telescopes show the faint bridge of light that connects the two galaxies.

To find M51, aim your telescope at Alkaid (Eta [η] Ursae Majoris), the "end" star in the handle of the Big Dipper. Make sure you're using a low-power eyepiece so the field of view is as large as possible; you can switch to higher powers once you've found M51. Carefully aim the telescope 2° east to the 5th-magnitude star 24 Canum Venaticorum. Now move the scope 1.5° south to a group of three 8th-magnitude stars that form a near right triangle. M51, which appears in finder scopes as a small circular haze, is located just east of the southernmost star in this triangle.

Once you've found M51, what you'll see depends on the telescope you're using. A 60mm refractor shows M51 as a tiny oval smear of gray light set against a sparse field of a dozen faint stars. With a 4-inch telescope, however, you'll resolve M51 and NGC 5195 and see a double glow. At moderate or high powers a 4-inch scope shows a slight central brightening in M51.

A 6-inch scope at 100x shows M51 as a large, oval patch of light with a smaller, circular glow, NGC 5195, just northwest of the large glow. In addition to the soft glow of the galaxies, the field of view contains one 8th-magnitude star 35' east-northeast of M51. This star

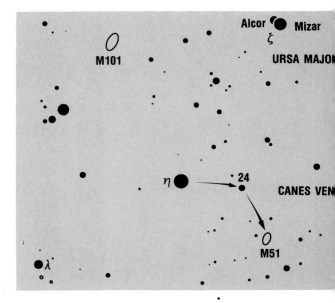

A 6-INCH SCOPE is required to see spiral structure in M51. On very dark nights 8-inch telescopes reveal a clear spiral pattern and a "bridge" of light linking M51 with NGC 5195.

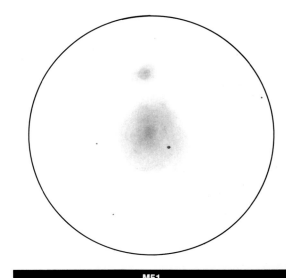

M51
8-inch f/10 SCT   100x   by David J. Eicher

M51
Magnitude 8.4   11.0' by 7.8'   by Tony Hallas and Daphne Mount

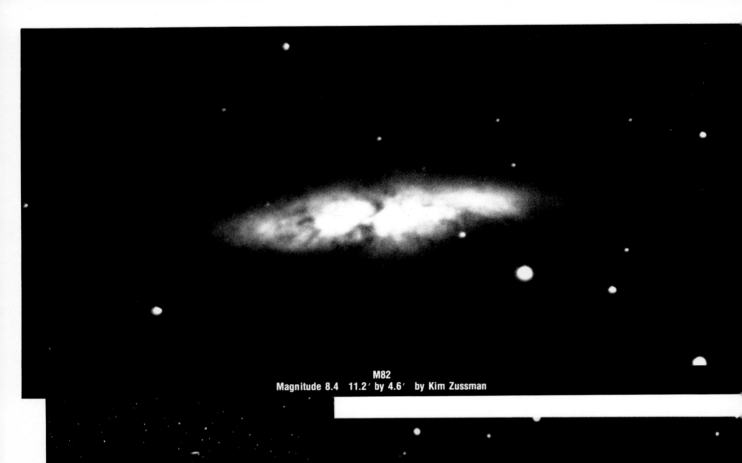

M82
Magnitude 8.4   11.2′ by 4.6′   by Kim Zussman

M81 and M82
by Mace Hooley

THE BRIGHTEST galaxy pair in the sky is M81 and M82 in northern Ursa Major. Small telescopes show a double patch of light: M81 is oval-shaped and its neighbor distinctly flattened.

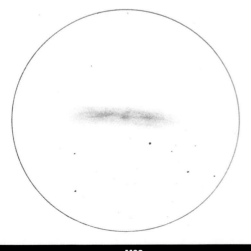

M82
8-inch f/10 SCT   100x   by David J. Eicher

M81
Magnitude 7.0   25.7′ by 14.1′   by Kim Zussman

is the southernmost of the three that form the right triangle. Midway between the galaxy and star is a small, distinctive triangle of 11th- and 12th-magnitude stars. Apart from this, the field is rather uniformly scattered with two dozen faint, white stars.

A good 6-inch or 8-inch telescope provides sufficient light-gathering power to show significant detail in the Whirlpool Galaxy. With high magnification, the first thing you'll notice is that the disk of M51 is not uniformly illuminated. The galaxy's core is significantly brighter than the glow that surrounds it. The surrounding glow appears mottled, or unevenly lit. On nights when the atmosphere is particularly transparent, you'll see the mottling isn't random but rather that the galaxy's disk is composed of spiral arms. This spiral pattern can be glimpsed with a 6-inch scope on the best of nights.

M51 in large amateur telescopes shows a wealth of detail not discernible with small scopes. A 12-inch telescope reveals a tiny, brilliant, starlike point of light in the galaxy's center. This pinpoint core lies within a 2'-diameter oval-shaped central glow. Surrounding the central glow are the galaxy's spiral arms, plainly visible on steady nights.

M51 contains two primary arms. One begins at the southern end of the central glow and winds clockwise, smoothly disappearing against the background sky. The other arm begins at the northern end of the central glow and winds clockwise, connecting to NGC 5195. The arms are clearly delineated by dark lanes; NGC 5195 also has dark patches that give it a mottled appearance in large telescopes. A solitary star dimly glowing at 12th magnitude lies 2' southwest of M51's center. This star is occasionally mistaken for a supernova in M51, but actually lies within our Galaxy, hundreds of thousands of times closer than M51 itself.

No matter what size telescope you have, two observing tricks will help you see detail in M51 and other galaxies. First, give yourself a minimum of twenty minutes in total darkness before observing so your eyes will perform at maximum sensitivity. This dilates your pupils and allows you to see faint light you would not see immediately after exposure to inside lighting. Second, use averted vision. Centering a galaxy in your field of view and glancing off to the edge of the field uses your eye's rods, the most sensitive receptors. With this technique you will often see details in a galaxy that are not visible when you look directly at the galaxy.

After M51, the finest galaxy in the spring sky is **M81** (NGC 3031) in Ursa Major. This bright spiral forms a wide pair with **M82** (NGC 3034), a peculiar galaxy oriented edge-on to our line of sight. To find the M81-M82 pair, aim your telescope at Dubhe (Alpha [α] Ursae Majoris), the end star in the bowl of the Big Dipper. Sweep 12° west and 2° north to 23 Ursae Majoris, a 4th-magnitude double star. Next, go 7° north to a 1°-long row of three stars, the brightest of which is 24 Ursae Majoris. From this point move your telescope 2° southeast to M81 and M82.

Once located, M81 and M82 are visible together in the same low-power telescopic field. Although M81 is brighter than M51, it is not nearly as detailed. A 4-inch telescope shows M81 as an oval-shaped nebulosity some 5' across and M82 as a smaller and much fainter sliver of gray light. Six-inch telescopes reveal a tiny, bright nucleus in M81 and a mottled appearance across cigar-shaped M82 (this is caused by dust lanes that cross the

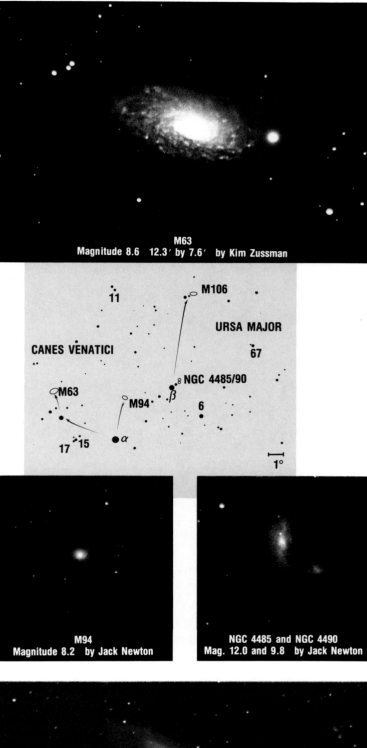

FIVE BRIGHT GALAXIES NESTLE in central Canes Venatici. M63 features spiral arms visible in large backyard telescopes. M94 has a tiny, bright nucleus. NGC 4485 and NGC 4490 form an interacting pair. M106 has arms visible in a 10-inch scope.

M108
Magnitude 10.1 8.3' by 2.5' by Martin C. Germano

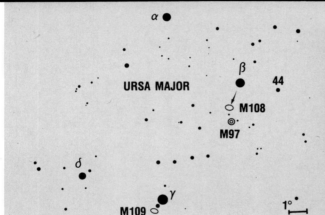

M109
Magnitude 9.8 7.6' by 4.9' by Martin C. Germano

SWIMMING IN THE BIG DIPPER'S BOWL are spiral galaxies M108 and M109. Small telescopes show M108 as a silver streak. M109 appears like an oval nebulosity with a bright center.

Bright Springtime Galaxies

Object	R.A. (2000.0) Dec.	Mag.	Size	Tirion	Uranometria
M81	9h55.6m +69°04'	7.0	25.7' by 14.1'	1, 2	I: 23
M82	9h55.8m +69°41'	8.4	11.2' by 4.6'	1, 2	I: 23
M108	11h11.5m +55°40'	10.1	8.3' by 2.5'	2, 6	I: 46
M109	11h57.6m +53°23'	9.8	7.6' by 4.9'	2, 6	I: 47
M106	12h19.0m +47°18'	8.3	18.2' by 7.9'	2, 6	I: 74
NGC 4485	12h30.5m +41°42'	12.0	2.4' by 1.7'	7	I: 75
NGC 4490	12h30.6m +41°38'	9.8	5.9' by 3.1'	7	I: 75
M94	12h50.9m +41°07'	8.2	11.0' by 9.1'	7	I: 75
M63	13h15.8m +42°02'	8.6	12.3' by 7.6'	7	I: 76
M51	13h29.9m +47°12'	8.4	11.0' by 7.8'	7	I: 76
NGC 5195	13h30.0m +47°16'	9.6	5.4' by 4.3'	7	I: 76
M101	14h03.2m +54°21'	7.7	26.9' by 26.3'	2, 7	I: 49
M102	15h06.5m +55°46'	10.0	5.2' by 2.3'	2	I: 50
NGC 5907	15h15.9m +56°19'	10.4	12.3' by 1.8'	2	I: 50

Object: Designation in Messier or NGC catalog. **R.A. and Dec.:** Right ascension and declination in equinox 2000.0. **Magnitude:** V magnitude of galaxy. **Size:** Diameter in arcminutes. **Tirion:** Chart on which object appears in *Sky Atlas 2000.0* by Wil Tirion. **Uranometria:** Chart on which object appears in *Uranometria 2000.0* by Wil Tirion, Barry Rappaport, and George Lovi.

galaxy). Unfortunately M81's spiral arms have an extremely low surface brightness and require a 10-inch or larger telescope to spot even with averted vision.

Two bright galaxies lie near the bowl of the Big Dipper. **M108** (NGC 3556) lies 2° southeast of Merak (Beta [β] Ursae Majoris), glows at 10th magnitude, and is an edge-on spiral galaxy. Visible in small scopes as a thin object resembling M82, M108 is a treat in large backyard telescopes. As with M82, we see a great deal of dark, obscuring clouds in M108 because of the galaxy's edge-on orientation. A 16-inch telescope shows small dark patches in the gray-green light of this galaxy. What appears to be a pinpoint nucleus in M108 is actually a dim foreground star.

M109 (NGC 3992) is a magnificent barred spiral galaxy glowing at 10th magnitude and lying within the same low-power field as Phecda (Gamma [γ] Ursae Majoris). Although this galaxy's spiral structure isn't as apparent as M51's, a large backyard scope will clearly reveal it on an especially dark night. M109 is not quite face-on, and its arms have a relatively low surface brightness. Small scopes show an elongated bright center in this galaxy surrounded by a dim fuzz. A 12-inch

M101
Magnitude 7.7 26.9' by 26.3' by Paul Roques

scope at moderate power reveals the galaxy's bar and its faint spiral arms.

A group of contrasting galaxies lies south of the Big Dipper. **M106** (NGC 4258) can be found next to a 6th-magnitude star some 5° east of Chi (χ) Ursae Majoris. This large, bright spiral is visible in 2-inch telescopes. An odd pair of interacting galaxies, **NGC 4485** and **NGC 4490**, lies 1° northwest of the bright star Beta (β) Canum Venaticorum. The brighter of the two is NGC 4490, an irregular barred galaxy glowing at 10th magnitude. NGC 4485, a much smaller irregular galaxy, is a 12th-magnitude object. Large telescopes show these galaxies as odd-shaped blobs of light that nearly touch.

Three-and-a-half degrees east of Beta lies **M94** (NGC 4736), a face-on spiral with an intensely bright center and extremely faint, diffuse outer arms. Most backyard scopes show only M94's nuclear area; even large scopes show the arms as a faint cloud surrounding the bright center. **M63** (NGC 5055) is a magnitude 8.6 spiral with knotty arms located 5° east of M94. M63 appears like an oval smudge of nebulosity in a 2-inch telescope. In a 6-inch scope the galaxy appears like a small, bright central glow with an extremely faint oval haze surround-

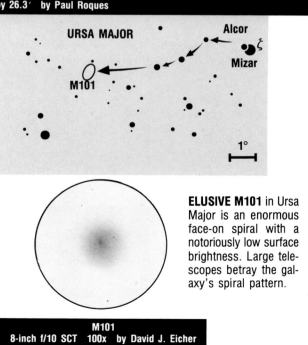

ELUSIVE M101 in Ursa Major is an enormous face-on spiral with a notoriously low surface brightness. Large telescopes betray the galaxy's spiral pattern.

M101
8-inch f/10 SCT 100x by David J. Eicher

NGC 5907
Magnitude 10.4   12.3' by 1.8'   by Kim Zussman

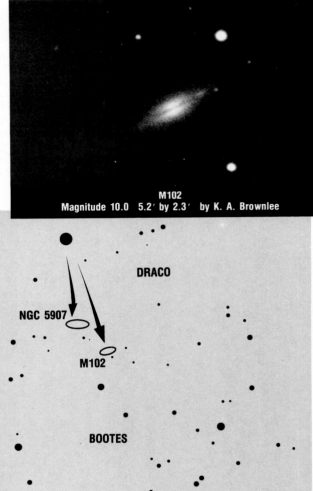

M102
Magnitude 10.0   5.2' by 2.3'   by K. A. Brownlee

A SLIVER OF LIGHT, edge-on NGC 5907 is one of the thinnest galaxies visible in small telescopes. Its neighbor M102 is a bright, spindle-shaped galaxy with a featureless disk.

ing it. A 10-inch or larger telescope reveals a faint, mottled spiral pattern in the galaxy's hazy outer envelope.

M101 (NGC 5457) provides one of the best test runs for star hopping in the entire sky. To find this big face-on galaxy, aim your scope at the bright double star Mizar (Zeta [ζ] Ursae Majoris). Move from Mizar through Alcor, its northerly companion, a distance of 2° to the 5th-magnitude star 81 Ursae Majoris. Now move east along a line consisting of 83, 84, and 86 Ursae Majoris. Just east of 86 lies a north-south line of 8th-magnitude stars. M101 lies 1° northeast of the center star in this line.

Although it is big, bright, and oriented face-on, M101 is notoriously difficult to observe because it has a low surface brightness. Observe this galaxy on nights when the sky is at its darkest and use low magnifications. A 6-inch scope shows a diffuse glow at the galaxy's position; a 10-inch scope demonstrates the galaxy's center is brighter than its outer nebulosity.

Another pair of galaxies is that composed of **M102** (NGC 5866) and **NGC 5907** in Draco. M102 is a lens-shaped galaxy glowing at 10th magnitude and measuring 5.2' by 2.3'. Although it is bright, M102's appear-

ance changes little from telescope to telescope. More exciting is its neighbor NGC 5907, located 2° northeast, one of the sky's best edge-on spirals. Even a 2-inch telescope shows this as a thin streak of light. Large backyard scopes reveal a thin dust lane bisecting the galaxy lengthwise.

Each of the galaxies we have surveyed is bright because it is close to us in space. Among the nearest galaxies beyond our own galaxy group are M81 and M82, which lie approximately 10 million light-years away. Although this sounds like an enormous distance, it is in our backyard as galaxies go. M51 is barely twice as distant, lying some 25 million light-years off. The most distant galaxy we've looked at is M109, which lies nearly 50 million light-years distant. That amateur astronomers can detect the faint barred spiral arms of M109 is testimony to the quality of backyard telescopes and the skill of amateur observers. □

These objects are best visible in the autumn evening sky.

The Whirlpool Galaxy
photo by Tony Hallas and Daphne Mount

Legendary Faint Galaxies

These galaxies are so faint, distant, or heavily obscured that they'll test the observing skills of even the most determined observer.
by Steve Lucas

Have you ever seen a *really* faint galaxy? If not, you're missing one of the great challenges of backyard observing. If you have, you know that seeing a faint galaxy tests your visual acuity, sharpens your observing skills, and rewards you with a glimpse of a very distant or low surface brightness object that few on this little planet have observed firsthand.

Galaxies are challenging targets for backyard telescopes because they are dim objects. Some galaxies are extremely distant and therefore appear dim — many are not much brighter than the sky background itself. Others are near us in space but are intrinsically dim, generating a small fraction of the light output of a normal spiral galaxy. Still others are bright and relatively close, but their light is obscured because they lie behind the disk of the Milky Way.

NGC 891 is the best all-around challenging galaxy in the winter sky. Located 3.5° east of the star Gamma Andromedae, NGC 891 is a magnitude 10.0 spiral galaxy that measures 13.5' by 2.8' in extent. It is an edge-on galaxy with a broad dust lane that splits it in half. Because of its inclination, dust lane, and large size, NGC 891 has an extremely low surface brightness. Therefore NGC 891 appears much fainter — and is trickier to find and observe — than its magnitude suggests.

To find NGC 891, insert a low-power eyepiece in your scope. This provides a wide field of view and a reasonably contrasty image of the galaxy. Aim the scope at Alpha Andromedae, the star that marks the northeastern corner of the Great Square of Pegasus. Follow the southern arc of bright stars that forms Andromeda by moving from Delta to Beta and on to Gamma Andromedae. From Gamma, move 3.5° east. In your finder scope you should see a double star with 6th- and 7th-magnitude components. NGC 891 lies 1° north of this double star.

Once you locate NGC 891, what you'll see depends on the telescope you are using. The galaxy is a challenge to see at all with a 2-inch or 3-inch telescope and shows a sliver of pale light oriented northeast-southwest that measures 6' across and lies in a field strewn with faint stars.

Six-inch telescopes easily show NGC 891, but with this size instrument the challenge is spotting detail in the galaxy. On all but the poorest nights a 6-inch scope reveals the galaxy's broad dust lane. A 6-inch scope at 100x shows the galaxy's central hub — the brightest part of NGC 891 — that measures some 3' long. The disk of NGC 891 is visible as a faint extension some 3' long on

TO FIND NGC 891, follow the stars of Andromeda from Alpha Andromedae, the northeast corner of the Great Square of Pegasus, to a point 3.5° east of Gamma Andromedae.

both sides of the galaxy's center.

Large telescopes show more detail and provide a brighter overall image of NGC 891. They also crowd the field with groups of faint stars. In a 12-inch telescope the central bulge of NGC 891 appears reasonably bright and is surrounded by an amorphous glow of light that smoothly tapers down to sharp points on each end. At high power the dust lane's edges appear slightly rough, and a faint star (approximately magnitude 12) shows up near each end of the galaxy, embedded in the galaxy's light.

When you observe NGC 891, take advantage of a few tricks of the trade to see as much as you possibly can. After locating the galaxy, switch to a medium-power eyepiece (about 15x per inch of telescope aperture). This will provide a dark-sky background, which is an ideal background for seeing faint details in the galaxy. Move your eye around the field of view, looking directly at the galaxy and also off to an edge of the field: this lets you use the part of your retina that is most sensitive to faint light. Also, move the telescope ever so slowly in a circular pattern while keeping the galaxy somewhere in the field. Human vision is sensitive to motion, so this movement could help you see detail in the galaxy.

After viewing NGC 891, move on to Cassiopeia's **IC**

NGC 891
Mag. 10.0   13.5′ by 2.8′   by Ron Potter and Jack Marling

DIM DWARF GALAXIES IC 10 and Leo I are challenging targets for small telescope users because the objects are inherently faint. NGC 891 is a low surface brightness edge-on galaxy with a hard-to-glimpse dust lane.

IC 10
Mag. 10.3   5.1′ by 4.3′   by Lick Observatory

Leo I
Mag. 9.8   10.7′ by 8.3′   by Martin C. Germano

Maffei 1
Mag. 11.4 14.6′ by Martin C. Germano

HEAVILY OBSCURED
Maffei 1 would be one of the brightest galaxies if it didn't lie directly in the thick of the Milky Way. Instead it is a challenge for the largest amateur telescopes.

Challenging Winter Galaxies

Object	R.A.(2000.0) Dec.		Mag.	Size	Tirion	Uranometria
IC 10	0h20.4m	+59°18′	10.3	5.1′ by 4.3′	—	I: 35
NGC 891	2h22.6m	+42°21′	10.0	13.5′ by 2.8′	1, 4	I: 62
Maffei 1	2h36.3m	+59°39′	11.4	14.6′	—	I: 38
Leo I	10h08.4m	+12°18′	9.8	10.7′ by 8.3′	13	I: 189

Object: Designation in NGC, IC, or other catalog. **R.A. and Dec.:** Right ascension and declination in equinox 2000.0. **Mag.:** V magnitude of galaxy. **Size:** Diameter in arcminutes. **Tirion:** Chart on which object appears in *Sky Atlas 2000.0* by Wil Tirion. **Uranometria:** Chart on which object appears in *Uranometria 2000.0* by Wil Tirion, Barry Rappaport, and George Lovi.

10, an even more challenging object. IC 10 is an irregular galaxy that lies on the outskirts of our Local Group of galaxies. Although it is close to home in an extragalactic sense, IC 10 is difficult to observe because it is intrinsically much fainter than a galaxy like NGC 891 and is obscured by the rich Milky Way in Cassiopeia.

IC 10 is easy to find: simply move your scope 2° east of Beta Cassiopeiae, the westernmost star in the "W" of Cassiopeia. Once you locate the galaxy's position, switch to a low-power eyepiece and look for a small "spot" of faint, milky light. In a 10-inch telescope IC 10 appears as a faint round glow, not more than 1′ across, lying in a field containing TV Cassiopeiae, which varies between magnitudes 7.2 and 8.2.

Still more challenging is the low surface brightness galaxy **Leo I**. Leo I is a dwarf elliptical galaxy in our Local Group. It is intrinsically faint and small, and we see it only because it lies reasonably close. This galaxy is especially challenging because it lies 20′ north of Regulus, one of the brightest stars in the sky. The glare from Regulus overwhelms the faint light from Leo I, which makes sighting the galaxy extremely challenging even with a large-aperture telescope.

To see Leo I, try using at least a 10-inch telescope. Aim your finder scope at Regulus and move the telescope a couple of degrees north. Insert a high-power eyepiece in the telescope, one that provides a 15′ field of view. Slowly move the telescope south toward Regulus looking for a faint glow, but make sure to stop short of reaching Regulus. If you don't, you'll flood your eye with photons and have to wait fifteen minutes to regain your dark adaptation.

Another method of observing Leo I is placing an occulting bar in front of Regulus, blocking the star's light but allowing light from faint stars — and Leo I — to reach your eye. To do this, cut a small piece of opaque material and tape it over one edge of the eyepiece lens. When you observe Leo I, make sure that light from Regulus is blocked by the bar.

The most challenging galaxy in the winter sky is **Maffei 1**, a heavily obscured object buried behind the Cassiopeia Milky Way. Discovered by Italian astronomer Paolo Maffei in 1968, Maffei 1 is a large spiral galaxy on the outer fringes of our Local Group — reasonably close but blocked by gas clouds and dust in our own Galaxy. If it were located in a barren region of sky, Maffei 1 would be one of the finest galaxies in the sky.

However, it isn't. Wedged between the Double Cluster in Perseus and the big bright clusters IC 1805 and IC 1848, the position of Maffei 1 is not difficult to find. Start by moving 2° northeast from the Double Cluster to a group of four 7th-magnitude stars, three forming an isosceles triangle. From there move 1.5° north to a double composed of 7th- and 9th-magnitude stars. Now move 3/4° northeast to a wide pair of 8th-magnitude stars. (The easternmost is a double.) Maffei 1 lies 15′ south of this wide pair.

Confusion reigns when it comes to sighting Maffei 1. Although amateurs can photograph this galaxy with long exposures, visual sightings are extremely difficult even with large apertures because the galaxy is really faint. Moreover, a little clump of very faint stars lies practically on top of the galaxy's center. In all likelihood many amateur sightings of Maffei 1 are really sightings of this clump of stars.

Observing these faint galaxies in the winter sky will challenge you in quite different ways. The overall low surface brightness and elusive detail of NGC 891 is within range of most amateur telescopes on cold, clear nights — all you need is persistence and a careful eye. IC 10 heightens the challenge of seeing an extremely faint object. Leo I, fainter still than IC 10 and adjacent to one of the sky's brightest stars, tests the most experienced and best-equipped backyard observers. And Maffei 1 lies behind a sprinkling of faint stars, which forces you to make sure you're seeing the galaxy and not unresolved light from distant suns.

As you warm up indoors and contemplate successes or failures, you can dream up new strategies — among them occulting bars, averted vision, and slow sweeping — to help pull in photons from some of the toughest deep-sky objects in the sky. □

These objects are best visible in the autumn evening sky.

Andromeda's NGC 891
photo by Bill Iburg

The Galaxies of Cetus

Cetus the Whale is chock full of pretty, challenging galaxies for small telescopes.

by David Higgins

Few constellations offer a window into the deep like the one in Cetus the Whale. Although this constellation appears rather empty when viewed with the eye alone, with a telescope Cetus comes alive with deep-sky wonders. By looking toward Cetus you are peering through a window that extends beyond our galaxy that reveals the realm of distant galaxies beyond.

Cetus covers much of the rather empty-looking autumn sky. Although it is not one of the brighter constellations, Cetus covers a total of 1,231 square degrees of sky and offers ample room for galaxies ranging from the extremely faint and difficult to the very bright and prominent. Some of the fainter galaxies require a large telescope, but there are enough galaxies within reach of the average amateur telescope to provide you with plenty of interesting deep-sky objects to observe on those clear, crisp fall evenings.

By far the brightest and largest galaxy in Cetus is **M77** (NGC 1068), the only Messier object contained within the whale's borders. This should give you some idea as to the apparent brightness of the other galaxies in this area of sky. Lying at a distance of 60 million light-years and glowing at magnitude 8.8, M77 is a member of an extraordinary class of objects known as Seyfert galaxies. These galaxies exhibit a distinctive emission line spectrum and have an active, highly energetic, starlike nucleus. Seyfert galaxies are also moderately strong radio sources.

M77 is the brightest of several galaxies clustered around the bright star Delta Ceti. It is a simple matter to find M77. Look about 1° southeast of Delta Ceti and you will find a softly glowing cloud with a very bright and condensed core. Measuring 6.9' by 5.9', M77 is a face-on Sb spiral galaxy. In general, face-on spirals are not exceptionally distinct compared with edge-on systems, but M77 is an exception.

In small telescopes M77 appears bright and has an intensely prominent nucleus. A noticeable field star that rivals the brightness of M77's core lies on the galaxy's southeast edge. Try different magnifications as you observe M77. Telescopes in the 8-inch range present a very different view. The nucleus appears much more stellar and the haze surrounding the nucleus is brighter and gradually fades outward. Mottling is visible, especially around the galaxy's southern edge. Some hint of spiral structure may be visible on good nights.

In 12-inch to 14-inch telescopes the view of M77's nucleus is improved. The nucleus is slightly more stellar

M77
Magnitude 8.8 Size 6.9' by 5.9' by Kim Zussman

M77
Magnitude 8.8 Size 6.9' by 5.9' by Jack Newton

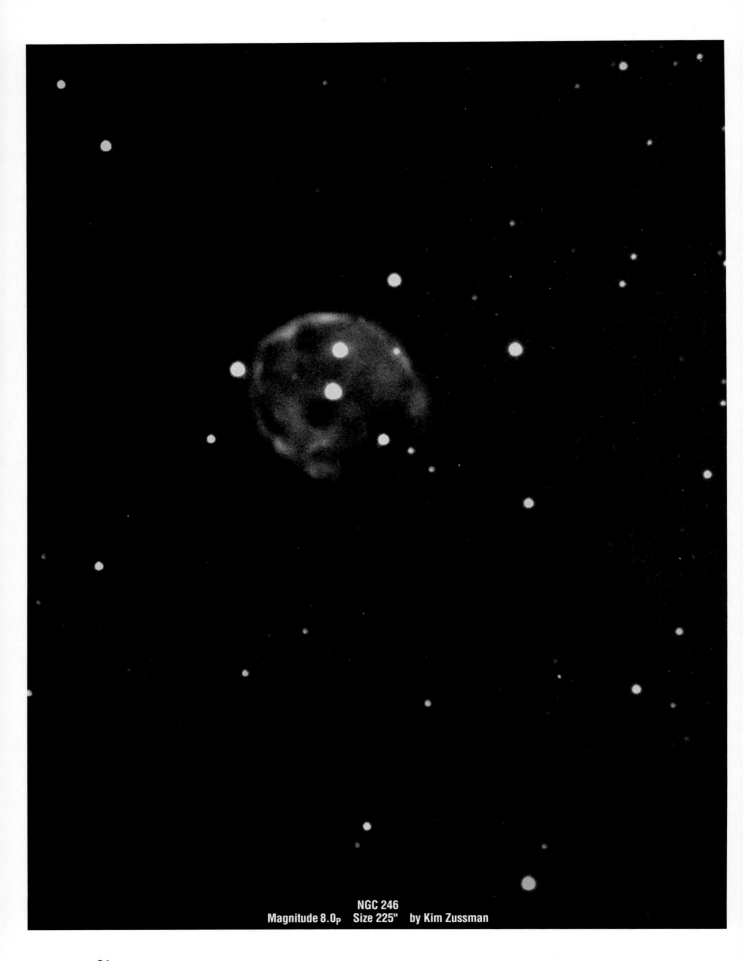

NGC 246
Magnitude 8.0p   Size 225"   by Kim Zussman

NGC 246 and NGC 255
Mags. 8.0p and 11.8 Sizes 225" and 3.1' by 2.8' by Mace Hooley

and the surrounding haze more intense. The galaxy's southern edge is much more mottled and pronounced, and on nights of good seeing the spiral structure becomes much easier to see. If you catch a really good night you can glimpse individual knots of nebulosity on the southern edge with averted vision.

Telescopes larger than 16 inches in aperture show an amazing amount of detail in M77. Spiral structure around the galaxy's southern edge is easily seen. The galaxy's nucleus appears absolutely starlike. The haze that extends outward from the nucleus is very bright and well defined.

Observing galaxies requires patience, skill, and some time-tested techniques. Dark adaptation is something that all observers, beginners and old hands, know is critically important. Don't be reluctant to try high magnifications. Each jump in magnification will provide you with a different and sometimes surprising view. Using high magnifications not only yields larger images but also greater contrast between the galaxy and surrounding sky. Averted vision also helps bring out hidden details that might not be apparent with direct vision.

About 30' northeast of M77, within the same low-power eyepiece field, is the faint edge-on spiral **NGC 1055**. This galaxy is elongated east-west and shows little structure or detail, except in large scopes. Glowing at magnitude 10.6 and measuring 7.5' by 3', NGC 1055 is visible in 6-inch scopes as a faint haze surrounding a bright, elongated nucleus. In large scopes averted vision reveals a small amount of mottling with a dark lane bisecting the nucleus.

M77 and NGC 1055
Mags. 8.8 and 10.6 Sizes 6.9' by 5.9' and 7.6' by 3.0' by Jeffrey L. Jones

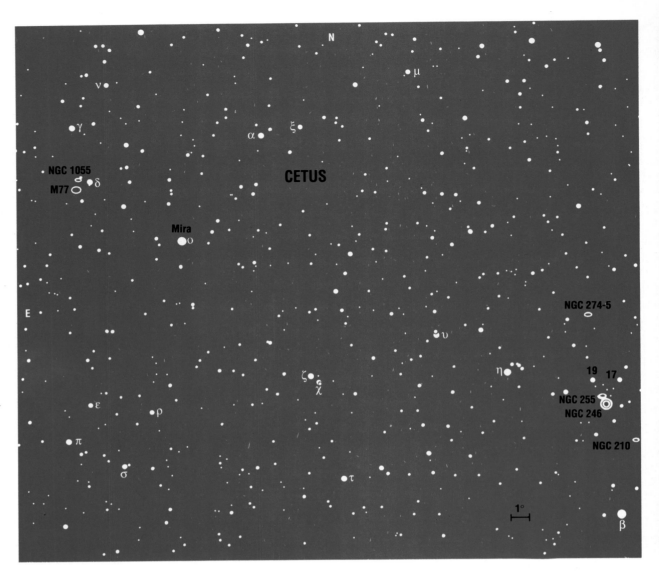

Galaxies in Cetus

Object	R.A. (2000.0)	Dec.	Mag.	Size
NGC 175	0h37.4m	-19°56'	12.1	2.6' by 2.4'
NGC 178	0h39.1m	-14°10'	12.6	2.0' by 1.2'
NGC 210	0h40.6m	-13°52'	10.8	5.4' by 3.7'
NGC 246	0h47.0m	-11°53'	8.0$_P$	225"
NGC 247	0h47.1m	-20°46'	8.8	20.0' by 7.4'
NGC 255	0h47.8m	-11°28'	11.8	3.1' by 2.8'
NGC 273	0h50.8m	-06°53'	13:P	2.6' by 0.9'
NGC 274	0h51.0m	-07°03'	12.9$_B$	1.7' by 1.6'
NGC 1055	2h41.8m	+00°26'	10.6	7.5' by 3.0'
M77	2h42.7m	-00°01'	8.8	6.9' by 5.9'

Object: Designation in Messier or NGC catalog. **R.A. and Dec.:** Right ascension and declination in equinox 2000.0. **Mag.:** V magnitude, B magnitude (subscript B), or photographic magnitude (subscript P). **Size:** Diameter in arcminutes or arcseconds.

Lying alongside the whale's head are several interesting galaxies awaiting inspection. One that is fairly bright but a little difficult to find is **NGC 210**. Located a little over 1° southwest of 18 Ceti, NGC 210 glows at magnitude 10.9, measures 5.4' by 3.7', and is an Sb-type spiral. In small scopes it is a large, faint oval of light with a nucleus a little brighter than the surrounding haze. In larger scopes the nucleus appears much brighter and extends north-south. Don't be afraid to use high magnifications on this galaxy.

Like M77, NGC 210 has a fainter neighbor that can be seen in the same low-power field. Lying about 20' southwest of NGC 210, **NGC 178** is a faint oval haze with no detail in small scopes. Ten-inch scopes show a brighter nucleus that extends a little north and south. The faint haze that surrounds the nucleus fades gradually outward.

Not too far northeast of NGC 210 lies another faint spiral. **NGC 255** is a small, face-on Sb spiral. Forming a lopsided triangle with 19 and 17 Ceti, NGC 255 lies about 1° south of these 5th-magnitude stars. Small scopes show a small oval haze with no sign of any nuclear brightening. Telescopes greater than 10 inches in aperture show the galaxy's nucleus as a bright ringlike

NGC 247
Magnitude 8.9   Size 20.0' by 7.4'   by Martin C. Germano

structure, surrounded by a faint glow.

NGC 255 shares the same low-power eyepiece field with **NGC 246**, a large planetary nebula that glows at photographic magnitude 8.0. Measuring 225" across, NGC 246 is easily seen with small telescopes when it is near the meridian. Low magnifications yield the best views; nebula filters like the UHC help greatly by boosting contrast. Several dark areas are visible in the nebula. The object's central star and several others inside the nebulosity are easily seen.

About 4° north of 19 Ceti you will find a little trio of galaxies with an unusual range of sizes and brightnesses. **NGC 274** is the largest of the three, measuring 1.7' by 1.6' and glowing at blue magnitude 12.9. Almost on top of NGC 274 and a little to the east is **NGC 275**, which glows at magnitude 12.5 and measures 1.5' by 1.2'.

The third member of the group is **NGC 273**, which lies just to the north. Measuring 2.6' by 0.9' and glowing at 13th magnitude, NGC 273 is the faintest of the three. Regardless of what size telescope you are using, there is little to be seen in this galaxy.

NGC 247 lies about 3° south of Beta Ceti, glows at magnitude 8.8, and measures 20' by 7.4'. NGC 247 is a very large galaxy. Its light is spread out over such a large area that its surface brightness is rather low. Small scopes show a faint haze without any central brightening and a 9th-magnitude star positioned in the southern portion of the galaxy.

Scopes in the 12.5-inch range show some hint of structure around NGC 247's edges. The northern portion is somewhat larger and more easily seen. The nucleus becomes visible with a gradual brightening outward. In large scopes the galaxy's nucleus is much larger and more extended. In a 17.5-inch scope the entire galaxy takes on a mottled appearance, with some individual knots of nebulosity taking shape around the edges.

Northwest of NGC 247 you will find a small, faint SBb spiral that requires some perseverance to find. **NGC 175** glows at magnitude 12.1 and measures 2.6' by 2.4'. This face-on barred spiral is not bright in small scopes, but it is worth finding. The simplest way to find NGC 175 is to draw an imaginary line between Beta Ceti and NGC 247. Halfway between the two extend a line westward about 2°. Small scopes will show a fairly large but faint haze with no apparent nucleus. Larger scopes will show a brighter middle with a hint of elongation north and south. The halo surrounding the nucleus gradually fades outward.

Keep in mind that the galaxies you are observing are millions of light-years away. In addition to their great distance, each one has its own complement of gas, dust, and stars. Spend some time with these far-away island universes. You will not only develop some highly useful skills, you will also enjoy peering outside of our galaxy into the great spaces beyond. □

These objects are best visible in the autumn evening sky.

Core of the Virgo Cluster
Anglo-Australian Telescope Board photo

Explore the Virgo Cluster

A small telescope is all you need to seek out numerous residents of the densest cluster of galaxies visible in our sky.

by Alan Goldstein

The constellation Virgo represents a comet hunter's nightmare. Nowhere can you observe a greater concentration of galaxies — big and small, bright and faint, elliptical and spiral. The 18th-century comet hunter Charles Messier cataloged no fewer than a dozen galaxies in central Virgo and six more in a 15° radius from that spot. However, although he found quite a few, Messier also missed plenty of galaxies.

Virgo contains so many galaxies the renowned astronomer Edwin Hubble called the area the Realm of the Galaxies. Hubble and his colleagues found that so many galaxies are visible in this area because in Virgo (and neighboring Coma Berenices) we are seeing the largest cluster of galaxies in our part of the universe. The so-called Coma-Virgo cluster is part of a supercluster, an enormous association of galaxy clusters, to which our own Milky Way is a distant member.

Spiral and elliptical galaxies dominate the Virgo Cluster. Several Virgo galaxies are huge — dwarfing our Milky Way, with its diameter of 100,000 light years. Consequently, you'll be amazed at how many galaxies you can see in this region, even with a pair of binoculars under dark skies. Telescopes in the 2.4- to 6-inch range reveal a large number of galaxies in Virgo. In some places a 6-incher shows five galaxies within a single field. Seven- to 10-inch scopes show even more galaxies. Telescopes 12-inches or larger in diameter are most effective at picking up the myriad of faint members of the Coma-Virgo cluster; these telescopes also show details such as arm structure in the brightest and largest Virgo galaxies.

Describing all galaxies in the Virgo Cluster visible with small telescopes is beyond the scope of this article. Instead, we'll examine only the best and brightest of them. You can find more by checking *Sky Catalogue 2000.0* or *Burnham's Celestial Handbook*.

Our first galaxy is **NGC 4216**. At magnitude 10.4, NGC 4216 is certainly bright enough to have been picked up by Charles Messier. He did not record it, but you can see it in a small telescope or good pair of binoculars under very good skies. It is a large Sb spiral galaxy seen in profile. The galaxy's nucleus, which is bright and condensed, has a bright star about 1' east of center. NGC 4216 is one of the best edge-on galaxies in the region, but is usually overshadowed by M104 to the south and NGC 4565 to the north.

There are many spiral galaxies in Virgo. One of

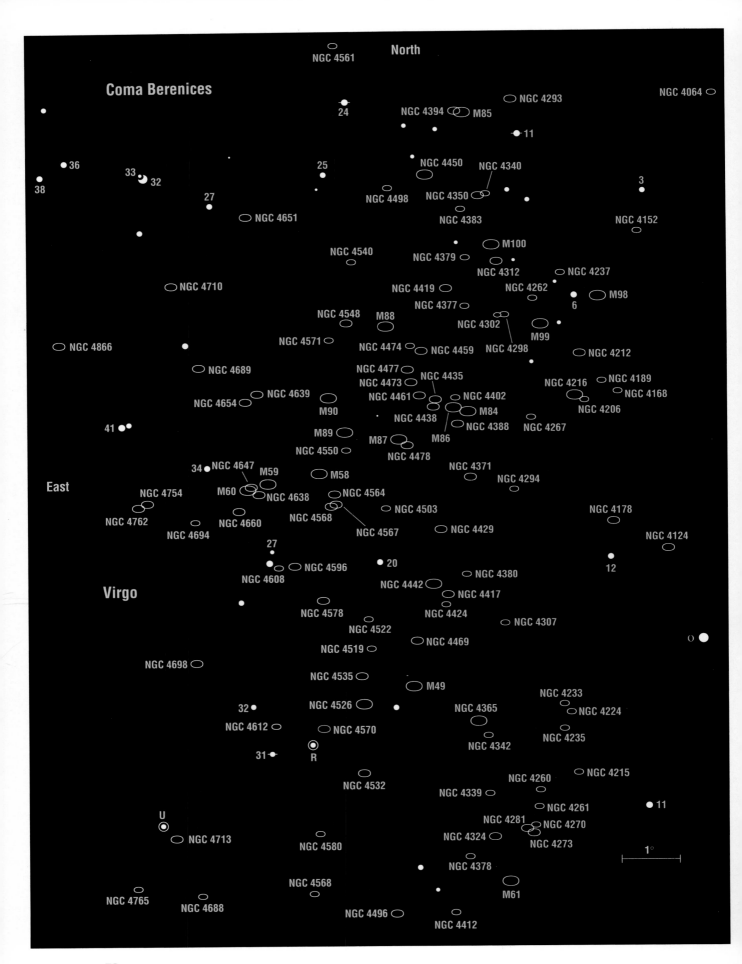

M61
Mag. 9.7   Size 6.0' by 5.5'   by Jack Newton

M89
Mag. 9.8   Size 4.2' by 4.2'   by Martin C. Germano

M90
Mag. 9.5   Size 9.5' by 4.7'   by Martin C. Germano

Bright Galaxies in Virgo

Object	R.A.(2000.0)	Dec.	Mag.	Size
NGC 4216	12h15.9m	+13°09'	10.0	8.3' by 2.2'
M61	12h21.9m	+ 4°28'	9.7	6.0' by 5.5'
NGC 4324	12h23.1m	+ 5°15'	12.6$_B$	2.5' by 1.2'
M84	12h25.1m	+12°53'	9.3	5.0' by 4.4'
M86	12h26.2m	+12°57'	13.0$_B$	7.4' by 5.5'
NGC 4402	12h26.1m	+13°07'	11.7	4.1' by 1.3'
NGC 4413	12h26.5m	+12°37'	13.0$_B$	2.5' by 1.7'
NGC 4425	12h27.2m	+12°44'	11.9	3.4' by 1.2'
NGC 4435	12h27.7m	+13°05'	10.9	3.0' by 1.9'
NGC 4438	12h27.8m	+13°01'	10.1	9.3' by 3.9'
M49	12h29.8m	+ 8°00'	8.4	8.9' by 7.4'
M87	12h30.8m	+12°24'	8.6	7.2' by 6.8'
NGC 4526	12h34.0m	+ 7°42'	9.6	7.2' by 2.3'
M89	12h35.7m	+12°33'	9.8	4.2' by 4.2'
NGC 4567	12h36.5m	+11°15'	11.3	3.0' by 2.1'
NGC 4568	12h36.6m	+11°14'	10.8	4.6' by 2.1'
M90	12h36.8m	+13°10'	9.5	9.5' by 4.7'
M58	12h37.7m	+11°49'	9.8	5.4' by 4.4'
M59	12h42.0m	+11°39'	9.8	5.1' by 3.4'
NGC 4638	12h42.8m	+11°26'	11.3	2.8' by 1.6'
M60	12h43.7m	+11°33'	8.8	7.2' by 6.2'
NGC 4647	12h43.5m	+11°35'	11.4	3.0' by 2.5'

Object: Designation in Messier or NGC catalog. **R.A. and Dec.:** Right ascension and declination in equinox 2000.0. **Mag.:** V magnitude of galaxy; subscript B denotes blue magnitude. **Size:** Diameter in arcminutes.

the best is **M61** (NGC 4303). This galaxy is a "transitional" barred spiral with characteristics of both a normal and a barred spiral. In a small telescope, this outlying member of the Coma-Virgo cluster is a pale oval measuring about 5' across. A moderate instrument under very good skies will reveal a bright nucleus and mottled patches of light from the unresolved spiral arms. In a 12-inch scope the hint of spiral structure becomes more apparent.

About 45' north of M61 lies a challenging galaxy for small telescope users. **NGC 4324** glows at 12.5 magnitude. This small, oval-shaped Sa-type spiral lies near the double star 17 Virginis. The star is somewhat distracting and makes the feeble glow of the galaxy hard to spot. A 6-inch or larger scope is normally required to spot this pretty galaxy.

M84, M86, NGC 4402, NGC 4435 and NGC 4438 create an interesting visual panorama. Between these five galaxies, all located in the same low-power field, we can see the combined light of at least a trillion stars! In addition, two much fainter edge-on spiral galaxies lie in the same field. **M84** is a type E1 elliptical galaxy. Photographically, it appears slightly oval. Visually, however, it appears virtually circular — especially in small instruments. M84 is strikingly bright: In large telescopes it shows no detail other than a bright nuclear region and a soft halo of light trailing off into the darkness. **M86** is slightly fainter than M84 and slightly more oval in shape. It is classified as an E2 elliptical galaxy. Some observers have noticed that the outer edges of M86 are fuzzier — less well defined — than those of M84. Compare the two in your telescope

Core of the Virgo Cluster
Photo by Kim Zussman

M87
Mag. 8.6   Size 7.2' by 6.8'   by Jack Newton

M49
Mag. 8.4   Size 8.9' by 7.4'   by Jack Newton

— do you see any difference? About 10' north of M86 is a faint edge-on spiral (type Sb), **NGC 4402**. This spindle-shaped galaxy has a low surface brightness. In smaller scopes, the elongated appearance is less evident, with the nuclear region being the most pronounced. In larger instruments, the fainter disk formed by the outer portion of the arms is visible. The equatorial lane requires a very large instrument to be seen.

South and slightly east of M86 lies **NGC 4413**, an SBb-type barred spiral. It has a low surface brightness, making it a difficult object for small telescopes. In moderate apertures, this galaxy forms a compact circular patch. Northeast lies **NGC 4425**, a faint SB0 type lenticular galaxy. It is fainter than NGC 4413, but is nonetheless visible in a 6-inch under very good skies. NGC 4425 appears as an oval haze with a bright nucleus.

NGC 4435 and **NGC 4438** form an unusual interacting pair. With small to moderate scopes, these galaxies appear as ovals. NGC 4435 and NGC 4438 are larger than they appear visually. NGC 4435, the northern member of this celestial "Siamese twin," is a lenticular galaxy of class SB0. The outer halo has a lower surface brightness than a comparable elliptical galaxy. NGC 4438 is a distorted, dusty spiral galaxy and is highly inclined to our line of sight. Consequently, you'll need a large scope to see detail in this pair of colliding galaxies. NGC 4438 has a lower surface brightness than NGC 4435 — perhaps from all of the dust — making it appear slightly fainter.

At magnitude 8.6, **M49** (NGC 4472) is another bright elliptical galaxy. Classed as an E4, this galaxy is more elongated in appearance than the other bright ellipticals save M59. Twelve-inch-or-larger telescopes bring out a bright nuclear region, but overall this galaxy brightens evenly from the center until fading sharply into the night sky. Smyth described this galaxy in 1844 as "a bright, round, and well-defined nebula; the nebula has a very pearly aspect." Another elliptical, **M87** (NGC 4486), is similar in appearance to M86 but is somewhat brighter. Classified as a type E0 peculiar elliptical, M87 appears as a round glow with a bright center. This galaxy, a strong radio and x-ray emitter known as Virgo A, has a small eruptive jet emanating from its nucleus.

At magnitude 10.9, **NGC 4526** is a bright galaxy of the S0/Sa type. This is a borderline galaxy, sharing characteristics of both lenticular (S0) galaxies and the Sa type, with a large central hub and small ill-defined outer arms. Some sources class this galaxy as an elliptical, but it is definitely a lenticular object. NGC 4526 is large and very oval in shape. With large scopes, it appears soft and milky. The poorly developed arms cannot be seen visually. A true elliptical, **M89** (NGC 4552), is classified as a type E0. It is reminiscent of M87, which lies a little more than a degree west but M89 is a little smaller and fainter. Some references call M89 a type E1, which would make it slightly elongated.

M90 (NGC 4569) is similar to and very slightly brighter than M61. It is considerably more elongated because it is more highly inclined to our line of sight. Located about a half-degree north and slightly east of M89, this spiral (type Sb) is an easy target in the smallest telescope or good binoculars. With small and moderate instruments, the inner portion of the hub of the galaxy is visible. The core is slightly brighter than the inner arms, and the outer arms have a much lower surface brightness. The nuclear region is distinct in 8-inch and larger scopes. The fine, delicate nature of M90's spiral arms make them difficult to observe. An uneven light in the hub is due to the

M 58
Mag. 9.8   Size 5.4' by 4.4'   by Rick Dilsizian

NGC 4567 and NGC 4568
Mags. 11.3 and 10.8   Sizes 3.0' by 2.1' and 4.6' by 2.1'
by Martin C. Germano

M 60 and NGC 4647
Mags. 8.8 and 11.4   Sizes 7.2' by 6.2 and 3.0' by 2.5'
by Martin C. Germano

large patches of dark nebulosity located there.

NGC 4567 and **NGC 4568** form an unusual pair of interacting spiral galaxies. As if they are physically connected, astronomers have named this system the Siamese Twins. Unlike their namesake, these distorted Sc type galaxies are not permanently attached, but are only cartwheeling past each other. In tens of millions of years, their passage will be complete. As they hurtle apart, they will probably be connected by an umbilical cord of stars, dust and gas for millions of years. In small telescopes this pair may seem to be a single, odd-shaped galaxy.

M58 (NGC 4579) is a bright spiral galaxy, equal in brightness to the giant M87. Like M90, this galaxy is a type Sb object. Its orientation is less inclined and the central hub is well exposed. The spiral arms, as is typical for Virgo galaxies, have a low surface brightness. Consequently, they are difficult to detect in small instruments and difficult to resolve in large ones. Photographs show a foreground star superimposed on one of the filamentary spiral arms. Can you see it? A 7th-magnitude star lies about 7' west of the galaxy. Place it out of the field of view to obtain the best view of M58. **M59** (NGC 4621) is a bright, compact glow about 1' east of M58. At 9.6 magnitude, it is readily visible in small instruments, even under light-polluted skies. M59 is an elongated elliptical galaxy of type E5. There are several other galaxies in the immediate area. **NGC 4638**, a 12.2 magnitude elliptical galaxy, lies to the southeast, forming a triangle with M60.

M60 (NGC 4649) is another large elliptical galaxy and, like M84, is classified as a type E1. M60 is one of the brightest in Virgo. In a small telescope, it appears as a roundish glow, brightening toward the center. Careful scrutiny with small instruments will reveal a second galaxy, **NGC 4647**. This galaxy is a spiral (type Sc) with a low surface brightness. It is often missed by observers looking at the well-known M60. The nuclear region of NGC 4647 is more condensed and is the easiest part to see with a small telescope.

The concentration of galaxies packed into Virgo provides backyard astronomers with plenty of fun galaxy searching. Instead of checking off one galaxy after another in a systematic fashion, consider some options. One project you may want to try is to examine the bright elliptical galaxies. Look at their classification type. Do they visually match what you consider the "proper" classification?

Scan the bright galaxies to determine which appear the brightest and how they vary in shape. You might be pleasantly surprised to see that the galaxies do not look like copies of each other. When you observe spirals, which have the lowest surface brightnesses? The central hubs of most are easy to see. But how many spiral galaxies reveal the soft outer glow from the arms? If you have a large instrument, especially 16-inches or larger, how many galaxies show resolved spiral arms? If you observe with a pair of binoculars or a good finderscope, how far beyond Messier's double handfull can *you* go? □

*These objects are best visible
in the spring evening sky.*

Elliptical Galaxy M87
photo by Jack Newton

Observing the Sculptor Group of Galaxies

Discover a little-observed yet bright and detailed collection
of spiral galaxies lying near the South Galactic Pole.

by Rod Pommier

NGC 55
Mag. 8.2$_B$ Size 32.4' by 6.5'   by Steve Quirk

NGC 247
Mag. 8.9   Size 20.0' by 7.4'   by Kim Zussman

Winter is a season that strikes fear into the hearts of deep-sky observers. Rather than visions of sugar-plum fairies, diehard galaxy fans see visions of icicles dancing through their heads, of the warmth and galaxies of spring that still lie a long way off. Yet if you're willing to dress warmly and brave the cold, a spectacular group of galaxies awaits your gaze in the evening winter sky. The Sculptor Group of galaxies includes NGC 55, NGC 253, NGC 300, and NGC 7793 in Sculptor, as well as NGC 247 in Cetus.

NGC 253 is the showpiece galaxy of the Sculptor Group, by far the largest and brightest member of the group. It is probably the easiest spiral galaxy to observe after the great spiral galaxy M31 in Andromeda. Although NGC 253 ranks as one of the great showpieces of the entire sky, it is rarely recognized as such.

NGC 253 is one of the best examples of an Sc-type galaxy. To find NGC 253, first locate the bright star Alpha Sculptoris. Then move 4.5° northwest, passing the fuzzy glow of globular cluster NGC 288, and you'll come upon NGC 253. In an 8- to 10-inch telescope, the view of NGC 253 is an impressive sight second only to that of M31. NGC 253 appears as a bright, elongated "silver dollar" rich in detail. The galaxy brightens gradually from the edges toward the central region, which contains a brighter, non-stellar core. Mottling is easily seen in the middle portions of the galaxy, where it surrounds the core. The mottling fades as it extends toward the periphery of the galaxy.

In a 12-inch or larger telescope, NGC 253 is stunningly bright. The central region appears distinctly non-stellar. The nuclear region also appears elongated along the major axis of the galaxy. The mottling is much more prominent as the nebulosity extends away from the center: dark lanes can be traced from the region surrounding the nucleus to the very edges of the galaxy. Regardless of the aperture used, NGC 253 is a beautiful sight that leaves an impression etched on your mind.

The next brightest galaxy in the Sculptor cloud is **NGC 55**, a large system seen nearly edge-on from our vantage point in the Milky Way. In contrast to other classic edge-on galaxies (such as M104, NGC 4565, NGC 891, or NGC 5907), NGC 55 exhibits neither a prominent central bulge nor an equatorial dust lane. Instead, large-scale photographs reveal large, irregular dust clouds and emission nebulae scattered across

NGC 253
Mag. 7.1   Size 25.1' by 7.4'   by Scott Hicks

the galaxy and a bright central mass displaced toward one end of the galaxy. For these reasons, NGC 55 is classified as an irregular galaxy and is thought to resemble the Large Magellanic Cloud in its structure. Gérard de Vaucouleurs believes the system to be a distorted, old, barred spiral galaxy with S-shaped spiral arms that are viewed nearly edge-on. The galaxy has a mass of about 46 billion solar masses and a luminosity of about 6 billion suns.

To locate NGC 55, first locate the 3rd-magnitude star Ankaa (Alpha Phoenicis). Many amateur astronomers in northern latitudes are unfamiliar with this star because it lies in the southern constellation Phoenix. However, Ankaa lies at the same declination as the southernmost stars in the tail of Scorpius, with which most amateur astronomers are very familiar. Therefore, be assured that if you can trace the tail of the scorpion along the southern horizon in summer, you'll also find Ankaa in winter. From there, move your scope 3° north to an east-west row of three evenly spaced 7th-magnitude stars. This row points directly west to NGC 55. From the westernmost star, proceed west about the same width as the row of three stars to arrive at NGC 55.

In an 8-inch telescope, NGC 55 appears large and elongated. Several of the features notable in photographs of this galaxy can be discerned. There is a bright nucleus which is clearly off-center, being displaced toward the northeast end of the galaxy. Mottling is evident in the central area of the galaxy.

Nearby lies the challenging galaxy **NGC 300**. This galaxy is interesting because it bears a striking resemblance to M33 in Triangulum. Both are visually large, class Sc spiral galaxies with compact nuclei and well-formed spiral arms that can be resolved into bright clumps of blue stars and emission nebulae on large-scale photographs. NGC 300 is inclined approximately 42° to our line of sight, similar to the 55° of inclination exhibited by M33. The distance to NGC 300 is approximately 6 million light-years. It has a mass of about 25 billion solar masses.

To locate NGC 300, start again with Ankaa.

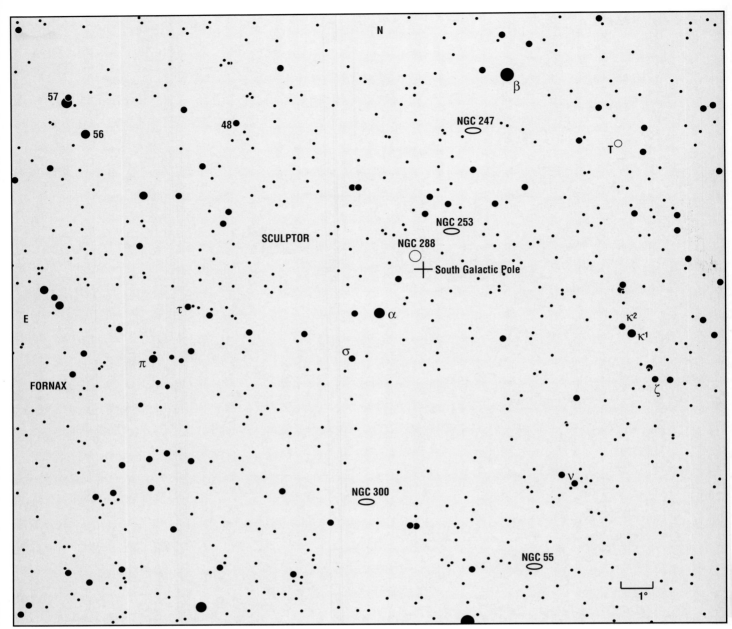

Bright Galaxies in Sculptor

Object	R.A. (2000.0)	Dec.	Mag.	Size
NGC 55	0h14.9m	−39°11'	8.2_B	32.4' by 6.5'
NGC 247	0h47.1m	−20°46'	8.9	20.0' by 7.4'
NGC 253	0h47.6m	−25°17'	7.1	25.1' by 7.4'
NGC 300	0h54.9m	−37°41'	8.7_B	20.0' by 14.8'

Object: Designation in NGC catalog. **R.A. and Dec.:** Right ascension and declination in equinox 2000.0. **Mag.:** V magnitude of galaxy; subscript B denotes blue magnitude. **Size:** Diameter in arcminutes.

Detail in NGC 253's spiral arms
Anglo-Australian Telescope Board photo

NGC 300
Mag. 8.7$_B$ Size 20.0' by 14.8'   Anglo-Australian Telescope Board photo

Move the telescope 5° northwest to a wide double star, Lambda¹ and Lambda² Sculptoris. A chain of 5th- to 8th-magnitude stars will lead you directly from Ankaa to Lambda¹ and Lambda². From Lambda², move the telescope ¾ of a degree north. Next, wait 11 minutes without moving the telescope (or using a clock drive), and Earth's rotation will carry NGC 300 into the center of the field of view. This technique is known as the star drift method. Use a low-power, wide-field eyepiece while waiting for the galaxy to appear and be sure to keep a very sharp eye out for it.

Through an 8- to 10-inch telescope, NGC 300 is a fairly difficult object because of low surface brightness. It has a bright, star-like nucleus which is much brighter than the nucleus of M33. The nucleus is surrounded by a large, tenuous, shell of nebulosity. This shell is very difficult to observe if there is much haze on the horizon. Therefore, a crystal clear night is essential to fully appreciate the extent of this galaxy. Averted vision helps to make the nebulous shell stand out, but no spiral arms or dust lanes can be seen.

The galaxy **NGC 247** lies in the constellation Cetus, yet belongs to the Sculptor Group of galaxies. It shares many characteristics with the previous two galaxies. Visually, it is like NGC 300 in that it is a large galaxy with a low surface brightness. Structurally, it is like NGC 55 in that it is ill-defined. Areas of light and dark, similar to those of NGC 55, are seen on large-scale photographs, but no definite spiral arms can be traced.

These galaxies are a mere sampling of what the Sculptor region has to offer. To explore some of the many fainter galaxies surrounding these objects, sweep the region with a low-power eyepiece and you'll surely come upon many faint smudges of light. As you log the details of each object, remember that each is an island universe consisting of billions of stars. And don't stay out too long, either — it's winter, after all! □

These objects are best visible in the autumn evening sky.

The Galaxies of Sextans

Discover a beautiful and neglected collection of spirals and ellipticals tucked away in springtime's sky-bound sextant.

by David Higgins

Lodged between the constellations of Leo the Lion and Hydra the Water Serpent resides one of the faintest constellations in the spring sky. The brightest star in Sextans the Sextant glows dimly to the naked eye, barely breaking the 4th-magnitude mark. Sextans holds a powerful secret, however. Although it sits in an apparently blank region of sky, it is absolutely loaded with galaxies for observers with small telescopes.

The largest and brightest galaxy in Sextans is **NGC 3115**. Located in the southern half of the constellation, it is not far from the Sextans-Hydra border. To find NGC 3115, start at the 4th-magnitude star Gamma Sextantis. From Gamma move about 6° eastward and you'll find the 5th-magnitude star Epsilon, or 22 Sextantis. NGC 3115 lies just north of a line drawn between the two stars. If you don't find the galaxy on your first attempt, sweep around a bit — NGC 3115 is bright enough to see easily in a wide-field eyepiece. This galaxy is a magnitude 9.2 edge-on spiral with a very high surface brightness. Consequently, you should feel free to use high magnifications as you examine NGC 3115. In telescopes 2 inches to 4 inches in aperture, the galaxy appears as a spindle of light 3' long, oriented northwest-southeast and with little or no brightening in its center.

In a 6-inch or 8-inch telescope, NGC 3115's spindle lengthens to about 4'. The galaxy's nucleus is much easier to differentiate from the haze that surrounds it. NGC 3115 gives the appearance of a smooth glow surrounding a bright core. Two relatively bright stars are visible in the same low-power field as NGC 3115, one on the northwest edge and the other on the southeastern tip. In really large backyard scopes, the nucleus becomes very bright while the core becomes almost stellar and takes on a triangular shape. With the help of large telescopes, the length of NGC 3115 grows to almost 5'. You won't detect any hint of dust lanes in this spiral — even though a large number of edge-on spirals have them, NGC 3115 doesn't.

An interesting group of galaxies lies in the north-central part of Sextans. Lying 4.5° northeast of Alpha Sextantis you'll find NGC 3169, NGC 3166, NGC 3165, and NGC 3156. **NGC 3166** is a magnitude 10.6 face-on spiral that measures 5.2' by 2.7'. In small scopes it appears as an oval haze with some central brightening and is elongated east-west. In scopes larger than 10 inches aperture, the galaxy's central region appears brighter and the core is nearly stellar. This galaxy handles high magnification quite well, so don't hesitate to crank up the power after you've found it. No dust lanes or structural details are visible.

NGC 3169 lies about 8' northeast of NGC 3166 and glows at magnitude 10.5. Measuring 4.8' by 3.2', NGC 3169 is easily visible in small telescopes. A 4-inch shows NGC 3169 as an oval haze elongated northeast-southwest. There is very little central brightening. An 8-inch scope reveals a bright nucleus and star-like center. A faint field star lies just off the eastern edge of the galaxy. Southwest of NGC 3166, you'll find another galaxy well suited for small scopes. **NGC 3165** lies 5' southwest of NGC 3166 and glows dimly at magnitude 14.5. Measuring 1.6' by 0.9' in extent, NGC 3165 is visible in small scopes only with good skies and high magnification.

In large backyard scopes, NGC 3165 is an elongated haze orientated north-south. There is some hint of central brightening, and only very large scopes show a stellar core. One other galaxy in the area is situated 30' southwest of NGC 3166. **NGC 3156** is a small, 13th-magnitude spiral that measures 2.1' by 1.2'. It is not as faint as NGC 3165, but isn't as bright or as large as NGC 3166. In a 4-inch scope, NGC 3156 is a small, oval haze that does not show any central brightening. Elongated roughly northwest-southeast, NGC 3156 has a bright nucleus, but overall, the galaxy has a smooth, evenly lit appearance. NGC 3156 is situated at the western edge of a bright triangle of stars, which makes finding the galaxy easy.

In the northeastern corner of Sextans, practically on the Leo-Sextans border, lies a bright, face-on spiral with a magnitude of 11.2. Although it measures 3.9' by 3.5', **NGC 3423** is a little harder to find than most of the Sextans galaxies. Start at 5th-magnitude 59 Leonis and, using a wide-field eyepiece, move about 2° west. If you don't find NGC 3423 after carefully sweeping the area, recenter on your original position 2° west of 59 Leonis, insert a higher magnification eyepiece, and look more closely. In small scopes, NGC 3423 is an oval haze elongated north-south. The galaxy's nucleus is brighter than the surrounding haze and several field stars dot the area. In a 10-inch or larger scope, the nucleus is strikingly bright and bar-like in shape, and a hint of bright areas within the galaxy's hazy envelope are visible.

NGC 2990 lies in the northwestern corner of

NGC 3115   Mag. 9.2   Size 8.3' by 3.2'
by Jeffrey L. Jones

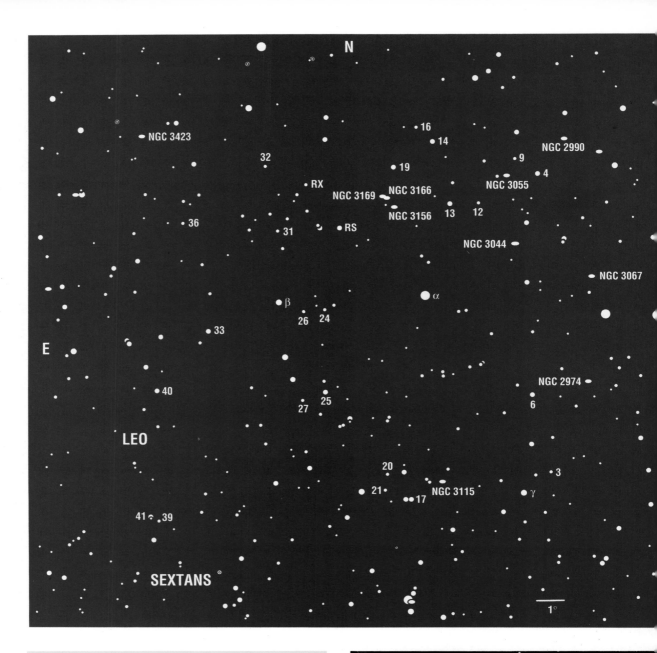

Galaxies in Sextans

Object	R.A.(2000.0)	Dec.	Mag.	Size
NGC 2967	9h42.1m	+0°20'	11.6	3.0' by 2.9'
NGC 2974	9h42.6m	-3°42'	10.8	3.4' by 2.1'
NGC 2990	9h46.3m	+5°43'	12.8	1.3' by 0.8'
NGC 3044	9h53.7m	+1°35'	12.0	4.8' by 0.9'
NGC 3055	9h55.3m	+4°16'	12.1	2.2' by 1.4'
NGC 3115	10h05.2m	-7°43'	9.2	8.3' by 3.2'
NGC 3156	10h12.7m	+3°08'	12.8$_B$	2.1' by 1.2'
NGC 3165	10h13.5m	+3°23'	14.5$_P$	1.6' by 0.9'
NGC 3166	10h13.8m	+3°26'	10.6	5.2' by 2.7'
NGC 3169	10h14.2m	+3°28'	10.5	4.8' by 3.2'
NGC 3423	10h51.2m	+5°50'	11.2	3.9' by 3.5'

Object: Designation in NGC catalog. **R.A.** and **Dec.:** Coordinates in equinox 2000.0. **Mag.:** V magnitude of galaxy; subscript B denotes blue magnitude and subscript P denotes photographic magnitude. **Size:** Diameter in arcminutes.

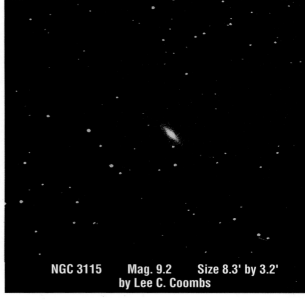

NGC 3115 Mag. 9.2 Size 8.3' by 3.2'
by Lee C. Coombs

NGC 3166 Galaxy Group
by Preston S. Justis

Sextans. This small, magnitude 12.8 face-on spiral is much more difficult to see than the previous galaxies. Measuring 1.3' by 0.8', this galaxy's low surface brightness makes it a tough target for telescopes less than 8 inches in aperture. The easiest way to find NGC 2990 is to start by locating the stars 25 Sextantis and 2 Sextantis. NGC 2990 lies just east of these stars and forms a lopsided triangle with them. Once you locate NGC 2990, expect to see a faint, oval haze with very little central brightening. A faint field star lies northeast of the galaxy. In large scopes, the nucleus appears brighter than the surrounding haze.

Southeast of NGC 2990 lies **NGC 3055**. Brighter and a little larger than NGC 2990, NGC 3055 is another face-on spiral that glows at magnitude 12.1. Measuring 2.2' by 1.4', NGC 3055 is easier to find than NGC 2990. Draw an imaginary line from Alpha Sextantis to Omicron Leonis. NGC 3055 lies approximately midway between these stars. In small scopes this galaxy shows up as a small, concentrated haze. No hint of central brightening is visible. A faint field star lies on the southwestern edge of the galaxy. In large backyard scopes, NGC 3055 presents a different appearance. The galaxy appears larger and shows a relatively bright nucleus.

NGC 3044 lies south of NGC 3055. This edge-on spiral is a much easier target than NGC 3055. It appears at first glance to be a smaller copy of NGC 3115. Not as bright or as large as NGC 3115, however, NGC 3044 glows meekly at magnitude 12 and spans 4.8' by 0.9'. Finding NGC 3044 is easy: Start at Alpha Sextantis and draw an imaginary line to the 5th-magnitude star 2 Sextantis. The galaxy lies midway between these two stars. Because NGC 3044 has a relatively low surface brightness, small scope users should employ moderately high magnifications to increase the contrast when observing this spiral.

NGC 2967 is a face-on spiral that glows at magnitude 11.6 and measures 3.0' by 2.9'. To find NGC 2967, start with Iota Hydrae and move about 1.5° northeast. In small scopes this object appears as a faint glow with some central brightening. In larger scopes the nucleus is brighter and easier to distinguish from the haze surrounding it. Overall, NGC 2967 is fairly concentrated toward its center but does not show a star-like nucleus. Now return to Iota Hydrae and use it as a starting point to find **NGC 2974**. This galaxy lies 3° southeast of Iota Hydrae. About the same size as NGC 2967, NGC 2974 is brighter and has a higher surface brightness. With a magnitude of 10.8 and dimensions of 3.4' by 2.1' it is visible in the low-power eyepiece of small scopes.

Sextans is not a bright constellation, nor does it have large, bright galaxies that make it a favorite of most observers. But the galaxies that do reside in Sextans will surprise you. Spend the time and make the effort to find some of the hidden galaxies of Sextans — you'll get valuable practice in star hopping techniques, the opportunity to observe little-known objects, and a keener eye for critically viewing these distant islands of stars. □

These objects are best visible in the spring evening sky.

Galaxy Hunting in the Great Bear

The region of the Big Dipper holds some of the brightest
and prettiest spiral galaxies in the sky.

by Richard W. Jakiel

Richly laden with galaxies, Ursa Major is one of the best hunting grounds for deep-sky objects. The Great Bear ranks fifth among constellations in numbers of deep-sky objects, with more than five hundred bright objects. Nearly all of these are galaxies and many are visible even in small telescopes. If you look at only the showpiece objects such as M81, M82, M101, M108, and M109, you'll miss galaxies even more spectacular than the so-called showpieces. With an 8-inch or 10-inch scope and a little planning, 150 to 200 galaxies are visible within the boundaries of Ursa Major. Many of the best and most interesting objects are not found among the brightest handful.

Most of these galaxies are part of the "Ursa Major Cloud," an extension of the Virgo Supercluster of galaxies, to which our own Local Group belongs. Many of Ursa Major's galaxies are "late-type" spirals and barred spirals. Late-type spirals are distinguished by bright H II regions of ionized gas and OB associations of hot young stars along with prominent dust lanes. Many of these structures are visible in medium to large backyard scopes. Additionally, most of these galaxies are highly inclined to our line of sight. In fact, nearly a third of them appear edge-on from our perspective.

Perhaps the best place to start your tour is in the vicinity of Merak (Beta Ursae Majoris), the southernmost pointer star in the Big Dipper. About 1.5° southwest of this star is **M108** (NGC 3556), a spectacular edge-on galaxy. At magnitude 10.0, M108 is visible in any telescope. In my 8-inch f/7 scope, the galaxy looks like a fat cigar lying at the end of a chain of 7th- to 9th-magnitude stars. Several bright and dark areas are visible along the galaxy's main axis. The bright areas are regions of high density stars and gas, and the dark areas represent dust.

In my 20-inch scope at 175x, M108 spans better than one-third of the field. It is elongated about 4:1 in an east-west orientation. The galaxy's surface is unevenly bright, with several bright condensations separated by dark rifts and bars. A 12th-magnitude star located nearly "dead center" looks like the nucleus of the system, but is actually a foreground object. M108 has a high surface brightness, so don't be afraid to use medium to high magnification (10x to 20x per inch of telescopic aperture) to examine surface details.

Less than 45' southeast of M108 is M97, the Owl Nebula. This bright planetary nebula lies well within our own Galaxy, and offers a welcome change of pace

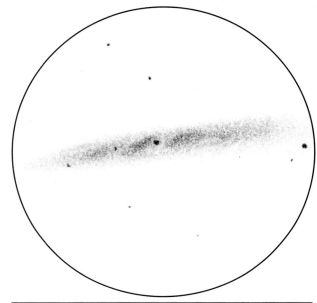

**M108   20-inch f/4.5 reflector at 175x
Sketch by Richard W. Jakiel**

during a night of galaxy observing. Forming a nice triangle with M108 and M97 is **NGC 3594**. This galaxy offers a good test of your observing skills because it glows dimly at 15th magnitude. In the 20-incher I saw only a faint, oval patch with a slight condensation.

About 2° southwest from NGC 3594, near the star 44 Ursae Majoris, is **NGC 3448**. It lies southeast of the star in the same low-power field. This object is an edge-on star system about half the size of M108 and about 1.5 magnitudes fainter. In the 8-inch scope, I saw a silvery streak about 2' by 0.5' that was weakly mottled and had a modest core region. In the 20-inch, the dimensions expanded to better than 4' by 1'. Overall, the texture is mottled. This object has a halo of rather low surface brightness, with several small, bright condensations spread along the galaxy's soft light.

About 2° southeast of M97 lies one of the brightest members of the Ursa Major Cloud, **NGC 3631**. This object is a large, face-on spiral with a total magnitude of 10.4. In the 8-inch scope I observed a small, bright core surrounded by a faint, diffuse outer halo. With the 20-inch scope I was able to discern a tiny, stellar nucleus

NGC 3877
Mag. 11.8   Size 5.4' by 1.5'   by Preston S. Justis

embedded within a mottled core region. The round halo was nearly 5' across. Very faint knots were also visible; these may be giant H II regions and OB associations, places where stars are forming and shine newly born.

Moving southward, the number of galaxies drops off rapidly. But there are a few bright galaxies within this sparse region. The most notable is **NGC 3675**, a bright, highly inclined spiral. It is an 11th-magnitude object located in an otherwise desolate part of the sky. In the 20-incher, the galaxy is oriented north-south and is elongated about 2.5:1. The core is oval and has a bright, tight nuclear condensation. On the east side of the core, an arclike dark lane of dust is clearly visible. Overall, I estimated the dimensions of the halo to be about 5' by 2' in the 20-inch scope. The halo has a high surface brightness.

Lying over 4° northeast is **NGC 3726**, a spectacular spiral galaxy. It is bright, at 10.4 magnitude, making it easy to see even in small scopes. This is one object that really has a visual impact in a large backyard scope. In my 8-inch scope I saw a large, oval object with a small, moderately dense core. It is aligned nearly north-south, and has a mottled surface texture. The true nature of the object is best revealed in a large scope, however. In the 20-incher at 175x, I saw a roughly ellipsoidal object with low-contrast spiral arms. I estimated the object measured about 6' by 4'. The northern arm is weak, with small knots outlining the structure. The southern arm is much more prominent, making a sharp turn to the east. Numerous H II regions help delineate the structure. The core is a short bar aligned with the long axis of the galaxy.

Just south of the bowl of the Big Dipper is a pair of interacting galaxies — **NGC 3718** and **NGC 3729**. NGC 3718 is easily the most peculiar bright object in this part of the sky. It is listed in Halton Arp's rogues' gallery, the *Atlas of Peculiar Galaxies*, as Arp 214; and this large, distorted barred spiral glows at 10.4 magnitude. With

M108
Mag. 10.0   Size 8.3' by 2.5'   by Gary Hug

NGC 3631
Mag. 10.4   Size 4.6' by 4.1'   by Preston S. Justis

NGC 3726
Mag. 10.4   Size 6.0' by 4.5'   by Preston S. Justis

my old 8-inch scope I didn't see much detail, only a large, nearly amorphous patch with a slight central condensation. When I peered through the 20-inch scope, the view was more fascinating. The first thing I noted was an unusual dust lane bisecting a nearly round disk. The dust lane was narrow near the nucleus and it gradually broadened away from the core. Overall, the dust lane resembled a huge "bow tie," similar to the appearance of the dust lane in NGC 5128 in Centaurus.

Lying in the same low-power field, NGC 3729 is a small, distorted galaxy. In the 8-incher all I could see was an elongated patch of light glowing at 11.5 magnitude and oriented nearly north-south. This object is definitely smaller and fainter than its neighbor. In the 20-inch scope it looked like an irregular oval of nebulosity. A low-contrast central bar formed the core of the object. With averted vision I could detect low-contrast irregularities along the axis of the bar.

About 1.5° due south of Phecda is the bright galaxy **NGC 3953**. This galaxy is a large, diffuse object oriented northeast-southwest and elongated about 2.5:1. In the smaller scope, I saw a bright stellar nucleus and a faint star in the western part of the halo. Switching back to the 20-inch scope, I observed low-contrast dust lanes near the oval core.

Much more interesting is the edge-on type S0 galaxy **NGC 4026**, an 11.7-magnitude object located only 1° northwest of NGC 4088. It has the classic "spindle form" typical of S0 galaxies. This shape is easy to see in a medium-sized scope, and in the 20-inch scope it looks a lot like its photographs. This galaxy has a high surface brightness. The "ends" taper down to points. Overall, this object looks like the classic example of a spindle galaxy, NGC 3115 in Sextans.

Nearly 1° south of the NGC 4088/4085 pair lies another nearly edge-on system. **NGC 4100** is an 11.6-magnitude object of moderate surface brightness. Through my 8-inch scope I saw an elongated silvery streak oriented almost north-south. Both the 8-inch and 20-inch scopes revealed a small tight core, but in the larger scope the core appeared somewhat off center.

Skimming southward close to the Canes Venatici-Ursa Major border are two more large galaxies. The largest of these is **NGC 4144**, an impressive, though dim, edge-on galaxy. Glowing at a meek 12th magnitude, it is the faintest large galaxy in the region. With the 20-inch scope, I noted an irregular, low-contrast dark lane close to the diffuse oval core. Not far away, **NGC 4096** is about the same size as the 4144 system, but 1.5 magnitudes brighter. It has a high surface brightness and is highly elongated. The surface is quite mottled and the oval core has a small stellar nucleus.

Dispersed around the 4th-magnitude star Chi Ursae Majoris are a number of bright galaxies. The most prominent of these is a large, low surface brightness spiral located in the same low-power field with Chi. **NGC 3877** is a medium-sized object glowing softly at magnitude 11.8. In my 8-inch scope, it looks like a fairly bright silvery streak with a moderately condensed core.

The last great galaxy group in the area lies near the double star 67 Ursae Majoris. About 2° northwest of the double you'll find **NGC 3938**, a very nice, magnitude 10.4 face-on spiral. When I viewed it with the 8-inch scope I saw a bright, round core surrounding a tiny nucleus. The galaxy's halo was large (about 4' across) and fairly faint. I did not view this object with a large scope, but photographs suggest that the spiral structure may be resolvable with a 17.5-incher.

Much closer to 67 Ursae Majoris is a pair of small galaxies, **IC 749** and **IC 750**. These galaxies are located close to a 8th-magnitude star about 1° west-southwest from 67 Ursae Majoris. IC 750 is the brighter at 11.8 magnitude, visible as a small nebulous streak with an irregular, "lumpy" core. In the 20-incher at 175x, I was able to resolve the core into three faint knots. IC 749 was barely visible in the smaller scope as a round, diffuse spot about 1' across. In the 20-incher, it was quite amorphous with a slight central condensation and the halo appeared to measure 2' across.

Our final objects, **NGC 4013** and **NGC 4051**, can be found north of 67 Ursae Majoris and west of NGC 3938. NGC 4013 is a thin, 12th-magnitude spindle of light. In the 8-inch scope it looks like a silvery needle with a slight central condensation. A 12th-magnitude star is superimposed near the center, masquerading as the galaxy's nucleus. NGC 4051 is a Seyfert galaxy, and like other members of this high-energy-output class it has an intensely bright starlike center. It is one of the brightest galaxies in the region, with a magnitude of 10.3. In medium-sized scopes, the galaxy's halo is quite elongated and the nucleus is easy to see. Using the 20-inch scope, NGC 4051 appears like a horizontally compressed "S," as short, faint arms are visible on both ends of the central bar. The nucleus is very sharp and bright, and the bar is rather smooth in texture.

Many other galaxies lie strewn across Ursa Major. And more are just across the border in Canes Venatici, such as those belonging to the NGC 4151 and M106 groups. The bowl of the dipper also contains many galaxies to explore. When you swing your telescope toward Ursa Major this spring, take a cursory glance at the super-bright galaxies you've seen many times. But then take some time to track down and study many of those lesser-known galaxies, whose special charm is unsurpassed by those in any region of the sky.

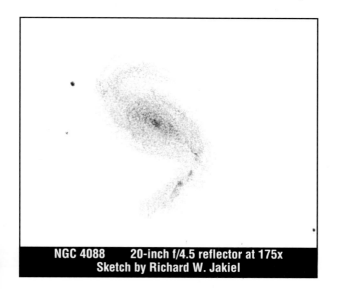

**NGC 4088   20-inch f/4.5 reflector at 175x
Sketch by Richard W. Jakiel**

These objects are best visible in the spring evening sky.

Galaxies of the Great Square

The Great Square of Pegasus holds some of autumn's most unusual and challenging galaxies for small telescope observers.

by David Higgins

The Great Square of Pegasus forms part of one of the most recognizable constellations in the autumn sky. Located far from the obscuring bands of dust and gas in the Milky Way, Pegasus is loaded with galaxies. Although not as well endowed with large or bright galaxies as its neighbor, Andromeda, Pegasus does provide a wide range of challenging objects to keep you busy on these cool nights of autumn.

Within the borders of the Square of Pegasus lie galaxies with some of the largest and smallest NGC numbers around. This results because the 0h line of right ascension passes through the constellation. Unfortunately these galaxies are not the brightest in the NGC nor are they the easiest to find. However, the galaxies of Pegasus offer an exceptional challenge for users of small telescopes. Spotting the galaxies within the Great Square will test your observing ability and the darkness of your sky. Can you find four of these galaxies? Six of them? Ten? The more of these distant spirals and ellipticals you record, the sharper your eye and keener your observing ability.

Gamma (γ) Pegasi anchors the southeastern portion of the Square and it is here that you will find the constellation's largest and brightest galaxy. **NGC 7814** is an edge-on spiral galaxy that measures 6.3' by 3.2' and glows at magnitude 10.5. This galaxy shows a great amount of detail in small scopes.

NGC 7814 is located slightly more than 3° northwest of Gamma Pegasi, just a couple of arcminutes southeast of a 7th-magnitude star. The galaxy's elongated halo is extended northwest-southeast. A 4-inch scope shows this halo as an oblong glow with no apparent brightening toward the center. The middle of the galaxy is a little thicker compared to the edges. You won't need averted vision to see this edge-on spiral so don't be afraid to increase the magnification.

In larger scopes NGC 7814's core appears condensed and lies just south of a faint field star. At high magnification you will see a slender dust lane that bisects the galaxy. The northwest and the southeast portions of the dust lane are the easiest to see; to spot the entire length of the dust lane requires a night of excellent seeing.

In 16-inch telescopes NGC 7814 takes on a different appearance. The dust lane becomes visible across its entire length and the core becomes much brighter. The faint field star to the north of the core is easily visible. Curiously, some observers have sighted the dust lane with scopes in the 4-inch to 6-inch range, while others with larger scopes have not mentioned seeing the lane. What you will be able to see depends on the size of the telescope you are using, how well it is collimated, your observing site, and the sky conditions.

About 1° south of Alpha (α) Andromedae lies a most interesting group of galaxies. One of the most attractive objects in the group is **NGC 1**, the first object in J. L. E. Dreyer's *New General Catalogue*. In a 4-inch scope NGC 1 appears like an oval spot glowing dimly at magnitude 13.4. This open-armed Sb-type spiral measures 1.9' by 1.5' and, like many of the galaxies in this region, has a rather low surface brightness. NGC 1's core is visible only with averted vision using scopes in the 6-inch to 8-inch range. Larger scopes provide a marginally better view, making the galaxy's core somewhat easier to disentangle from the outer envelope of haze.

Several arcminutes south of NGC 1 lies **NGC 2,** another of those small-number NGC galaxies that suffers from low surface brightness. NGC 2 measures 1.4'

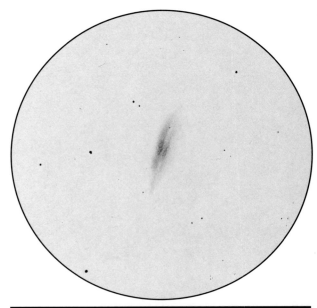

NGC 7814
Sketch by D. J. Eicher 17.5-inch f/4.5 reflector at 213x

NGC 7814
Mag. 10.5   Size 6.3' by 3.2'   by Kim Zussman

NGC 1, NGC 2, and NGC 16
Mags. 13.4$_p$, 14.8$_p$, and 12.0   Sizes 1.9' by 1.5', 1.4' by 0.9', and 2.1' by 1.2'   by Joe Liddell

by 0.9' and at 14.8 magnitude requires averted vision. In small scopes it is a faint oval haze with no central brightening. Larger scopes of course give a brighter view but do not provide much improvement overall.

About 30' east of NGC 1 and NGC 2 and 1.5° south of Alpha Andromedae lies the much brighter galaxy **NGC 16**. Glowing at magnitude 12.0 and measuring 2.1' by 1.2', NGC 16 is an easy target for small telescopes. A 6-inch instrument shows NGC 16 as an oval haze with no central brightening. In a 10-inch scope NGC 16 is oval but the core appears much brighter than in the 6-inch view. Large scopes show a stellar nucleus and a bright haze surrounding the core.

While you're in the area, drop about 2° southeast of NGC 1 and NGC 2. There you will find **NGC 23**, a 2.0' by 1.9' spiral that glows at magnitude 12.0. Much brighter than its northerly neighbors, NGC 23 is easily visible as an oval elongated north-south in a 6-inch scope. A 12-inch scope shows a stellar nucleus and crisply defined edges about the galaxy's perimeter. At first glance NGC 23 appears to have a double core, but on closer inspection you will see a faint field star located about 20" east of the galaxy's center. When you're finished with NGC 23, nudge your scope about 10' southeast and try **NGC 26**. This 14th-magnitude spiral is difficult for 6-inch scopes but telescopes in the 12-inch range reveal its oval shape.

About 1° north and a little west of 81 Pegasi lies a little trio of galaxies. **NGC 7769** is the brightest of the three, glowing at magnitude 12.1. This Sc-type galaxy measures 1.8' across and appears as a round disk in small scopes. Telescopes in the 12-inch range show the core brightening ever so slightly toward the center and the galaxy appears slightly elongated east-west. Scopes in the 17.5-inch range present a little different view. The core appears stellar and the haze that surrounds the nucleus is much better defined.

East-southeast of NGC 7769 you will find a 12.3-magnitude companion designated **NGC 7771**. This SBa-type spiral is almost as bright as NGC 7769 and measures 2.7' by 1.3', appearing somewhat extended northwest-southeast. In 6-inch scopes the core is brighter than the surrounding haze but little detail is visible. In 12-inch telescopes the core is slightly brighter and extended. **NGC 7770** is the last member of the trio. Situated south of NGC 7771, this 14.5-magnitude Sa-type spiral measures 1.0' by 0.9' and appears as a faint circular patch of light in a 12-inch scope at high power.

If you travel across the Square of Pegasus to its southwestern corner, you'll come upon Alpha Pegasi, the brightest star in Pegasus. Northwest of Alpha lie several faint galaxies. About a degree northwest of Alpha Pegasi you'll spot the Sc-type spiral **NGC 7448**. This 11.7-magnitude object is elongated southeast-northwest and measures 2.7' by 1.3'. In 6-inch and

NGC 1, NGC 2, and NGC 16
Mags. 13.4$_P$, 14.8$_P$, and 12.0   Sizes 1.9' by 1.5', 1.4' by 0.9', and 2.1' by 1.2'   by Bill Sorrells

NGC 7814
Mag. 10.5   Size 6.3' by 3.2'   by Bill Iburg

8-inch telescopes NGC 7448 is faint with no central brightening. With a 12-inch scope, the galaxy's oval shape is a bit more distinct and the core appears bright and somewhat mottled.

Some 30" east of NGC 7448 lies a fascinating group of galaxies. **NGC 7463**, **NGC 7464**, and **NGC 7465** are all faint in small telescopes but do provide an interesting challenge for users of small scopes. NGC 7465 is the brightest of the three, glowing at magnitude 13.3 and measuring 1.6' by 1.0'. This type SB0 barred spiral is oval in shape with no apparent nuclear brightening visible in small scopes. In 12-inch scopes the nucleus becomes brighter with a uniform haze surrounding it.

NGC 7464 is the faintest of the three and lies just to the west of NGC 7465. At magnitude 14.3, NGC 7464 appears very faint in 6-inch or 8-inch scopes. Measuring 0.7' across, this faint elliptical galaxy presents a faint round haze that is best seen with averted vision. Even in large backyard scopes there is no central brightening. NGC 7463 is just north of NGC 7464 and, although it is listed at magnitude 13.3, it is not as bright as NGC 7465. (I suspect one of the magnitudes is incorrect.) This faint Sbp galaxy is elongated east-west and measures 3.0' by 0.8'. In small scopes it does not show any central brightening and even larger scopes don't show much more than a uniform haze.

Beta (β) Pegasi marks the northwestern corner of the Great Square and serves as the starting point for our last three galaxies. About 6° southeast of Beta you will find the 4th-magnitude star Upsilon (υ) Pegasi. Aim your scope 30' east of Upsilon and you'll see a pair of contrasting galaxies that will challenge your perception skills. **NGC 7673** is the brighter and larger of the two. At magnitude 12.7, this type Sc peculiar spiral measures 1.7' by 1.6' and is elongated north-south. Telescopes in the 6-inch to 8-inch range show NGC 7673 as a small oval haze with very little central brightening. The challenging galaxy **NGC 7677** lies just a few arcminutes southeast of NGC 7673. Glowing at magnitude 13.9 and measuring 1.9' by 1.2', this galaxy has a very low surface brightness. Averted vision reveals a faint, slightly elongated haze with a small central condensation.

Move your telescope a little over 1° southeast of Upsilon and you'll see **NGC 7678**, an Sc-type galaxy that radiates dimly at magnitude 12.2 and measures

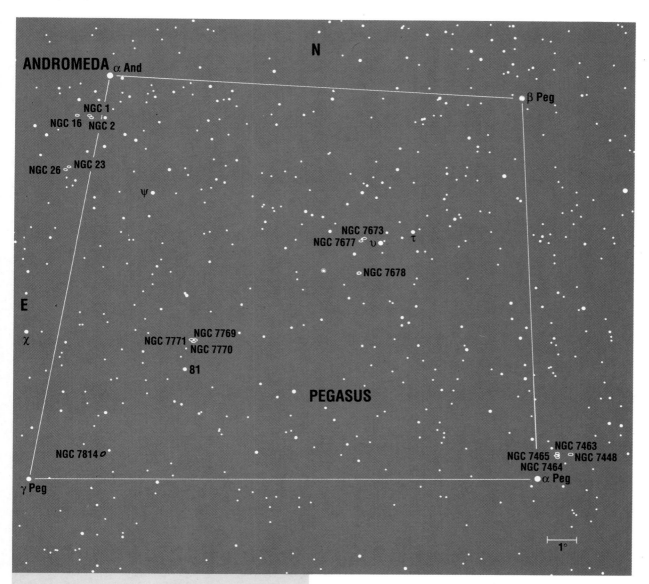

Galaxies in the Great Square

Object	R.A.(2000.0) Dec.	Mag.	Size
NGC 7814	0h03.3m +16°09'	10.5	6.3' by 3.2'
NGC 1	0h07.3m +27°43'	13.4_P	1.9' by 1.5'
NGC 2	0h07.3m +27°41'	14.8_P	1.4' by 0.9'
NGC 16	0h09.1m +27°44'	12.0	2.1' by 1.2'
NGC 23	0h09.9m +25°55'	12.0	2.0' by 1.9'
NGC 26	0h10.4m +25°50'	13.9_P	2.2' by 1.6'
NGC 7448	23h00.1m +15°59'	11.7	2.7' by 1.3'
NGC 7463	23h01.9m +15°59'	13.3_B	3.0' by 0.8'
NGC 7464	23h01.9m +15°58'	14.3_B	0.7' by 0.7'
NGC 7465	23h02.0m +15°58'	13.3_B	1.6' by 1.0'
NGC 7673	23h27.7m +23°35'	12.7	1.7' by 1.6'
NGC 7677	23h28.1m +23°32'	13.9_P	1.9' by 1.2'
NGC 7678	23h28.5m +22°25'	12.2	2.3' by 1.8'
NGC 7769	23h51.1m +20°09'	12.1	1.8' by 1.8'
NGC 7770	23h51.4m +20°06'	14.5_P	1.0' by 0.9'
NGC 7771	23h51.4m +20°07'	12.3	2.7' by 1.3'

Object: Designation in NGC catalog. **R.A. and Dec.:** Right ascension and declination in equinox 2000.0. **Mag.:** V magnitude; subscript P denotes photographic magnitude and subscript B denotes blue magnitude. **Size:** Diameter in arcminutes.

2.3' by 1.8'. NGC 7678 lies in a triangle of 11th-magnitude stars. Once you spot the little triangle, a keen look with averted vision will reveal the soft light of the galaxy. In 6-inch to 8-inch scopes NGC 7678 is an even glow with no central brightening. In 12-inch scopes the nucleus appears somewhat brighter than the surrounding nebulosity. In large backyard telescopes, those 16 inches and larger, NGC 7678 is slightly elliptical and its nucleus is much brighter than the haze surrounding it.

Autumn evenings provide a cool, comfortable time for galaxy hunting. During these convenient nights Pegasus and its Great Square ride high along the meridian and offer a unique and curious blend of galaxies both large and small, bright and challenging. Try finding a few of these remote denizens of the cosmos and you'll be exploring a region of space millions of light-years from home. □

These objects are best visible in the autumn evening sky.

Autumn's Galaxies: The Best and the Brightest

This is the perfect time of year to explore the sky's brightest galaxies.
by Max Radloff

Galaxies stir almost everyone's imagination. These vast, stately mansions range from dwarfs to giants and contain from a few million to hundreds of billions of stars. In shape, they can be elliptical, spiral, barred, or irregular, and we see them at any angle between edge-on and face-on. Each is a universe of stars by itself; and with a 6-inch or 8-inch scope you can explore not only their variety of shape, but also some of their internal components, such as star clouds, nebulae, dust lanes, and even globular clusters.

Fortunately, the autumn sky displays the best and brightest galaxies, conveniently placed for evening viewing. As if to make viewing galaxies ideal, cool autumn nights are typically free from summer's haze, presenting the year's best, most transparent evenings.

Autumn's greatest showpiece is the **Andromeda Galaxy**, M31 (NGC 224), the largest and brightest spiral galaxy in the sky. With a mass 300 billion times that of the Sun, the Andromeda Galaxy is one of the most massive galaxies in our local part of the universe. Its distance of 2.2 million light-years makes M31 the most remote object visible to the unaided eye.

On early star charts M31 appears as *nebulosa*, Latin for "misty or foggy." This description fits its naked-eye appearance well. In moderately dark skies the galaxy is visible as an elongated, hazy patch, slightly brighter in the center, and fading away on the edges until it blends into the sky background so gradually that you can't quite say when it disappears. Under average conditions the galaxy appears about 1° long, but its apparent length doubles when viewed under a truly dark sky. M31's appearance is quite sensitive to sky conditions. After you're familiar with it, a glance at the galaxy serves as a good indication of how dark and transparent the sky is.

In binoculars and small telescopes M31 is an impressive sight. The light from the galaxy's millions of stars combines into a smooth, milky glow that more than fills a low-power field. The galaxy's light brightens dramatically to a stellar nucleus. This rise in brightness from edge to center is even more striking at high powers.

Close observation of the entire galaxy reveals that M31's light does not glow uniformly. On the galaxy's northwest side, the light fades smoothly toward the edge, but it suddenly cuts off, reappears, cuts off again, and again reappears more faintly before finally fading completely. The light disappears behind two thin lanes of dust that encircle the galaxy, obscuring the light from stars lying behind them. If M31 is not too high in the

M32
Magnitude 8.2 Size 8' by 6'
by Mike Sisk

NGC 205
Magnitude 8.0 Size 17' by 10'
by Martin C. Germano

NGC 147
Magnitude 9.3 Size 13' by 8'
by Roger Sliva

NGC 185
Magnitude 9.2 Size 11' by 10'
by Martin C. Germano

THE BRIGHTEST SPIRAL GALAXY in the sky, M31 in Andromeda is visible to the naked eye as an oval patch of light. Telescopes reveal dust lanes and ghostly spiral arms.

M31
Magnitude 3.5 Size 178' by 63' by Tony Hallas and Daphne Mount

M33
Magnitude 5.7   Size 62' by 39'   by Tony Hallas and Daphne Mount

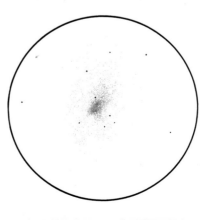

M33
17.5-inch f/4.5 Reflector at 71x
by David J. Eicher

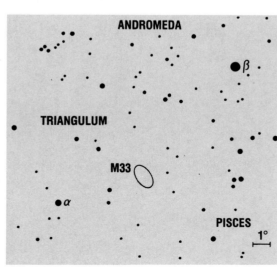

SPIRAL STRUCTURE is easily visible in galaxy M33 on a dark night. This galaxy contains one of the largest known nebulae.

M74
Magnitude 9.1   Size 10' by 9'   by Bill Iburg

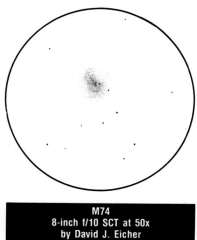

M74
8-inch f/10 SCT at 50x
by David J. Eicher

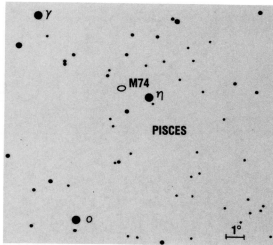

THE FACE-ON SPIRAL M74 is a challenge to find on less-than-dark nights. The galaxy's arms require averted vision and a clear, moonless sky.

eastern sky, take a look west and you can see this same phenomenon in our own Galaxy. Called the "Great Rift," it divides the Milky Way from Cygnus to Ophiuchus into two streams of light. Under excellent conditions, the dust lanes in M31 can be traced along much of the length of the galaxy and seem to have irregular edges. Under really black skies, the dust lanes give M31 a three-dimensional appearance.

About 1.5° southwest of the nucleus along the major axis of the galaxy lies an enormous star cloud. This object was prominent enough to nineteenth-century observers to receive its own NGC number, **NGC 206**. This object measures 4' by 2' and appears as a patch of light brighter than the rest of the galaxy that surrounds it. In physical terms this star cloud is a dense grouping of intrinsically bright stars that spans some 3,000 light-years. NGC 206 is visible in a 3-inch telescope under a dark sky. Observers with larger scopes may see much fainter patches of light along the galaxy, which mark the positions of dimmer star clouds.

With an 8-inch or larger telescope, look for a few of the brightest globular star clusters surrounding the Andromeda Galaxy. Because they glow at 13th magnitude, these globulars are challenging objects that require a detailed finder chart showing their exact locations. They also require enough patience to star hop through dense starfields to identify the clusters. If you want to take on the challenge, see the finder chart in the Autumn 1984 issue of *Deep Sky* magazine.

Beginning observers are sometimes disappointed when viewing galaxies, seeing only some feeble light with little apparent detail. One essential element for galaxy observing is a reasonably dark sky. If you live in a light-polluted area, you'll have to drive to a dark-sky site. Beyond that, galaxy observing requires only a little common sense, a few tricks, a measure of patience, and

M77
Magnitude 8.8   Size 7' by 6'   by Jack Newton

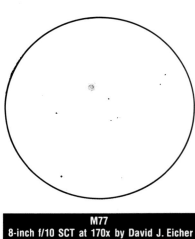

M77
8-inch f/10 SCT at 170x by David J. Eicher

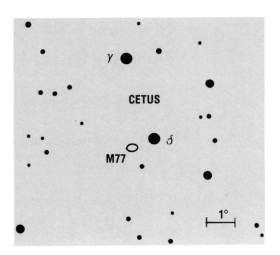

TINY M77 in Cetus is a tightly wound spiral with an intensely bright nucleus. Locate M77 by aiming 1° southeast of Delta Ceti.

experience.

A number of little factors will improve your galaxy viewing. Start with a well-collimated telescope with clean optics and eyepieces. Try to observe on the darkest, most transparent nights when the dew point and humidity are low. Drive a little farther than usual to reach the darkest site you can. Make sure that once you get there, you dark-adapt your eyes to maximum sensitivity. Although half an hour is usually considered sufficient, being in the dark for two or three hours will heighten your eye's sensitivity to faint light even more. Use averted vision — glancing to one side of the field — to see the most detail you can. Gently move or rock the telescope because your eye is especially sensitive to motion. This will help with spotting particularly faint details. And finally, don't be afraid to use high powers on faint objects like galaxies. The eye can see details in large, faint objects better than in small, bright ones.

The Andromeda Galaxy has eight known satellite galaxies. Four of these are visible in small telescopes, two of them within the same low-power field as M31 itself. (Four of the galaxies are beyond the reach of amateur instruments.) **M32** (NGC 221) is a small elliptical galaxy only about one hundredth the size of M31. It is visible in binoculars about 25' south of the nucleus of M31 as one of a pair of 8th-magnitude "stars." The point of light to the southwest is a star in our Galaxy and the one to the northeast is M32, thousands of times more distant and distinguishable by its fuzzy, nonstellar appearance. Through a telescope, M32 shows a bright nucleus with its light gradually fading toward the edges. Elliptical galaxies normally show no details through the eyepiece other than these gradations of light, although images from the largest telescopes show that M32 has dust lanes and a few small star-forming regions. M32's bright nucleus and high surface brightness make it well

NGC 253
Magnitude 7.1   Size 25' by 7'   by Bill Iburg

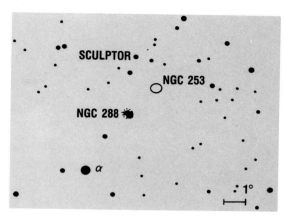

WONDER OF THE SOUTH, galaxy NGC 253 in Sculptor rides low on the horizon for Northern Hemisphere viewers. Nonetheless, the galaxy's huge size and brightness make it one of the sky's greatest attractions.

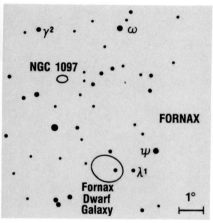

BARRED SPIRAL NGC 1097 in Fornax is visible in small telescopes as a circular patch of light surrounded by a mottled haze. On exceptionally dark nights the galaxy's bar is plainly visible at high power.

NGC 1097
Magnitude 9.3   Size 9' by 7'   by Martin C. Germano

worth examining at moderate and high powers.

The other companion galaxy near M31 is **NGC 205**, sometimes identified as "M110." This object is technically not included in the Messier catalog. Although Messier knew of this object in his lifetime, he learned about it after the final publication of his list. NGC 205 is located 40' northwest of the nucleus of M31. Like M32, it is a small elliptical galaxy, but NGC 205 is larger and far more elongated than M32. It is enlightening to compare the two galaxies under different sky conditions. Both produce nearly the same amount of light; but because NGC 205 is larger and its light more spread out, it appears dimmer than M32.

Two more satellite galaxies of M31 lie 7° away in the neighboring constellation Cassiopeia. Both are also small ellipticals, comparable in size to the star cloud NGC 206. The visual appearance of **NGC 185** is similar to that of NGC 205, only slightly smaller and fainter.

NGC 185 is not quite as elongated as NGC 205, but like NGC 205 it has a faint stellar nucleus. In a dark sky NGC 185 is visible with a small telescope, but even a moderate amount of sky brightness can make spotting it quite difficult.

Even fainter is **NGC 147**, another satellite galaxy lying just east of NGC 185. A 6-inch or larger telescope shows this galaxy; but because of its very low surface brightness, sky conditions are important. Look for NGC 147 on the darkest of nights and try all the observing tricks if you don't see it right away. NGC 147 appears as a faint, glowing patch with no detail.

Another member of the Local Group of galaxies provides the best opportunity to explore the spiral arms of a galaxy. **M33** (NGC 598), also called the Pinwheel Galaxy, is a nearby spiral that is much smaller than the Milky Way. Luckily for galaxy observers, this loose spiral is tilted about halfway between face-on and edge-on to our line of sight, allowing a good view of its arms. M33 is at the edge of naked-eye visibility, visible only as a tiny patch of fuzz on the darkest nights. In binoculars or a 2-inch telescope, however, M33 appears as a large, pale, ghostly oval about the size of the gibbous Moon. Because of its low surface brightness, many observers have difficulty locating M33. The key is to search from a dark-sky site using the widest field possible; this shows M33 as an impressive object, filling most of the field of view. From the galaxy's bright central area, two spiral arms extend outward in the shape of a backwards "S," and the entire area around them is filled with a diffuse glow.

About 12' northeast of M33's nucleus lies **NGC 604**, an enormous complex consisting of stars, gas, and dust. At 11th magnitude, this object is easily visible in small scopes even at a distance of 2.5 million light-years, and it forms a pair with a field star of similar magnitude. NGC 604 is similar to the Orion Nebula but is thirty times larger, measuring 1,000 light-years in diameter. Examining NGC 604 with high power and a nebula filter shows subtle striations and details in its structure.

If you observe M33 patiently, you can see other details. Using medium powers and gently moving the telescope's field will show you a few star clouds and dust lanes, and the two prominent spiral arms contain numerous bright patches. M33 has not produced any recorded supernovae, but its active star-forming regions, loaded with blue supergiants, make M33 the type of galaxy where a supernova is likely to appear.

Similar to M33 in a number of respects is **M74** (NGC 628) in Pisces, a face-on spiral with a low surface brightness. M74 can be difficult to find, and it is rated by many observers as one of the most challenging of the Messier objects. M74 is much smaller than M33 because it is twelve times as distant. Spotting M74's spiral arms is a challenging test of observing skill, but using moderate powers on a dark night will show them as subtle areas in the galaxy's outer glow.

M77 (NGC 1068) is the brightest of a small group of galaxies surrounding the bright star Delta (δ) Ceti. M77 is small but bright, and at a distance of about 60 million light-years it appears in binoculars as a 9th-magnitude "fuzzy star." Telescopically, M77 appears as an irregularly shaped patch of haze with a very bright nucleus. M77 is the closest of an exotic class of galaxies called Seyferts, named after the astronomer who discovered them, Carl Seyfert. Seyfert galaxies have anomalously bright and energetic nuclei and behave somewhat similar to quasars, although they are not nearly as powerful. Once you've located M77, switch to high power to examine the galaxy's intense nucleus and the bright areas around it. With a 12-inch or larger scope you may detect some of the galaxy's tightly wound spiral arms.

NGC 253 is a member of the Sculptor Group of galaxies, the closest group beyond our own Local Group. Distances to galaxies are still hotly debated, but a distance of about 6 million light-years is probably accurate for the Sculptor galaxies. NGC 253 is the brightest galaxy visible from northern latitudes not listed in the Messier catalog, and it is brighter than all but a handful of those observed by Messier. In binoculars NGC 253 appears as a prominent, tapering, dish-shaped glow over 20' long. Telescopes show NGC 253 as an impressive sight, revealing a bright, elongated oval nebulosity sim-

Great Autumn Galaxies

Object	R.A. (2000.0)	Dec.	Mag.	Size	Tirion	Uran.
NGC 147	0h33.2m	+48°30'	9.3	13' by 8'	4	I: 60
NGC 185	0h39.0m	+48°20'	9.2	11' by 10'	4	I: 60
NGC 205	0h40.0m	+41°41'	8.0	17' by 10'	4	I: 60
M32	0h42.7m	+40°52'	8.2	8' by 6'	4	I: 60
M31	0h42.7m	+41°16'	3.5	178' by 63'	4	I: 60
NGC 253	0h47.6m	−25°17'	7.1	25' by 7'	18	II: 306, 307
M33	1h33.9m	+30°39'	5.7	62' by 39'	4	I: 91
M74	1h36.7m	+15°47'	9.1	10' by 9'	10	I: 173
M77	2h42.7m	−00°01'	8.8	7' by 6'	10	I, II: 220
NGC 1097	2h46.3m	−30°17'	9.3	9' by 7'	18	II: 354

Object: Designation in Messier or NGC catalog. **R.A. and Dec.:** Right ascension and declination in equinox 2000.0. **Mag.:** V magnitude of galaxy. **Size:** Diameter. *Tirion:* Chart on which object appears in *Sky Atlas 2000.0* by Wil Tirion. *Uran.:* Chart on which object appears in *Uranometria 2000.0* by Wil Tirion, Barry Rappaport, and George Lovi.

ilar in appearance to but smaller than the Andromeda Galaxy. Because of the southern declination of NGC 253, northern observers have difficulty seeing detail in this galaxy; the only routinely visible details are a bright, ball-like nucleus, a bright central hub, and mottled light and dark areas along the galaxy's arms. On the other hand, Southern Hemisphere observers can see dust lanes, bright patches in the galaxy's arms, and patchy knots of dust.

The bright galaxy **NGC 1097** in Fornax is the only barred spiral in our survey. At a declination of −30°, northern observers may be tempted to pass by this object. However, NGC 1097 is clearly visible because of its size and brightness. In telescopes NGC 1097 has a bright, round nuclear glow surrounded by a diffuse, oblong halo of greenish light. Catch this galaxy when it's near the meridian in a 6-inch or 8-inch scope and you'll be able to spot its distinctive bar, which runs along the length of the galaxy. Just beyond the bar on the galaxy's northwest end lies a 13th-magnitude companion galaxy.

Difficult as it may be to comprehend the enormous sizes of galaxies, they are in turn parts of even larger structures in the universe. An enormous string of galaxy groups and clusters begins with the Sculptor Group on the southern horizon near the south galactic pole and contines through the Local Group to the Milky Way. This string of galaxies continues on the other side of the Local Group and stretches into groups of galaxies in Ursa Major and Canes Venatici. Still more groups follow along the northern horizon past the north galactic pole to the heart of the Virgo Cluster and beyond. Altogether, this association of galaxies spans at least 100 million light-years. Although it taxes the imagination to visualize a string of thousands of galaxies, each with many millions of stars, this is only a fragment of the observable universe. When you're under the stars enjoying the views of autumn's great galaxies, take a few moments to think about the immensity of it all. You'll feel just a little differently about the world after you go back inside. □

These objects are best visible in the autumn evening sky.

NGC 253 in Sculptor
photo by Bill Iburg

Standout Winter Star Clusters

With a small scope you can discover
a variety of star clusters that lie in the winter Milky Way.
by Alister Ling

The attraction of star clusters is simple. Star clusters are deep-sky objects every amateur astronomer or novice can enjoy. Unlike galaxies, bright star clusters pierce the cloak of light pollution in suburban skies, and because of this, star clusters are easy to observe with small telescopes even if the sky overhead is slightly hazy or somewhat bright.

Open star clusters are small groups of a few dozen to a few hundred stars that form from clouds of gas in the Milky Way's disk. Star clusters are fun to look at because no two are alike. Clumps, knots, and strings of material coalescing into stars create clusters with round or oblong shapes, irregular distributions of stars, and prominent chains or loops of stars stretching all the way across your telescope's field of view — distances that correspond to trillions of miles in space. Colors also play prominently: even with a telescope as small as 2 inches in aperture you will see stars of different colors in the brightest winter clusters.

M41 (NGC 2287) is the best open cluster in the winter sky. From a dark site, this magnitude 4.5 cluster is easily visible to the naked eye. The cluster's position just 4° south of Sirius — the brightest star in the sky — makes finding it a snap. Simply aim your finder scope at Sirius, then slowly sweep south, and you'll come upon a tightly packed group of bright stars.

M41 measures over 40' across — it's larger than the Full Moon! — so try observing it with a low-power eyepiece first. Relax and take a long, close look. Move your eye slowly around the field, carefully examining each little subgroup of stars that together forms the cluster. With such a wide field of view you can scan the entire cluster and the surrounding area all at once. Notice that even in a low-power eyepiece M41 appears large and round and holds several dozen bright stars.

If you're observing with a 4-inch or larger telescope, you'll see a scattered group of three dozen stars set against a blue-black backdrop. Although the cluster is generally circular, individual stars group into threes and fours within the cluster. A glance at the brightest star in M41, a 7th-magnitude gem located in the cluster's center, shows it is noticeably orange. Of the remaining dozen brightest stars, four others are orange and the others intensely blue-white. Do they all look white to you? Try putting the stars ever-so-slightly out of focus. This will help you see the colors more easily; they'll appear as colored fringes around the bright central star.

Unusual patterns of stars surround M41's central star. Shift your gaze east of the cluster's center and you'll see a concentration of bright blue, white, and orange stars. Just a few arcminutes northeast of the central star, two orange stars glowing at 8th and 9th magnitudes form the center of a C-shaped ring of six stars with its open end oriented northeast. Eight-inch or larger scopes show eight fainter stars in this figure, including a faint double comprised of twin white stars. Twenty arcminutes southeast of M41's center lies a 6th-magnitude blue star that is not a member of the cluster but simply a foreground object.

Several arcminutes southwest of the cluster's center lies a pair of 8th- and 9th-magnitude stars. The brighter

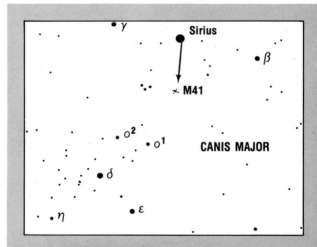

M93
Mag. 6.2 22'
by Jack Newton

SPARKLING M41 is located 4° south of Sirius, the sky's brightest star. A long-exposure photo (opposite page) shows the cluster's several dozen members. M93 in Puppis contains as many stars as M41 but is more condensed.

M41
Mag. 4.5 38′ by Paul Maley

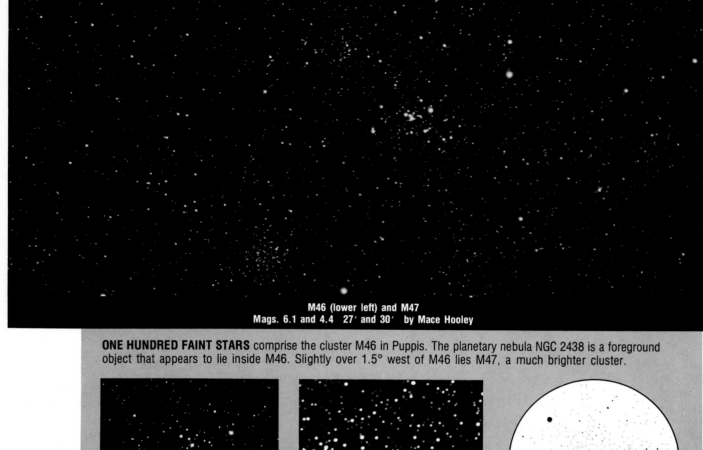

M46 (lower left) and M47
Mags. 6.1 and 4.4 27' and 30' by Mace Hooley

ONE HUNDRED FAINT STARS comprise the cluster M46 in Puppis. The planetary nebula NGC 2438 is a foreground object that appears to lie inside M46. Slightly over 1.5° west of M46 lies M47, a much brighter cluster.

M47
Mag. 4.4 30' by Jack Newton

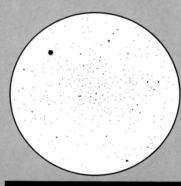

NGC 2438
Mag. 10.1 66" by Jack Marling

M46
8-inch f/10 SCT 50x by David J. Eicher

star has an orange tint and its companion is pale blue. This pair lies opposite the C-shaped ring relative to the central star. Slightly farther to the southwest a little row of four 11th-magnitude stars marks one of the easiest features to identify in M41. North of the double star you'll see a squashed triangle of stars readily visible in 4-inch instruments. North of the cluster's center lies a large clump of bright, blue-white stars — this area contributes significantly to the cluster's overall magnitude.

Scouting around the edges of the cluster, you'll notice how quickly the bright stars thin out in number. Now center the group and look about halfway out to the edge of the field of view, paying attention to the central area. This technique, called averted vision, is one of the most valuable tools in visual astronomy. Looking off to one side exposes the most light-sensitive part of your eye to the center of the cluster. Try it a few times — many "extra" stars will become visible and fill the cluster with faint stardust.

If you observe M41 on a number of nights, you'll notice it is satisfying even when the Moon is up or the night offers less-than-perfect seeing or transparency. This is the case with other clusters as well. Of course these objects look best when the atmosphere is steady and the Moon is out of the sky, but star clusters are observable just about anytime.

Unlike many small deep-sky objects, large open clusters like M41 are best seen in small or medium telescopes with fields of view large enough to include the entire cluster and some surrounding sky. Very large telescopes with their relatively small fields of view let you see only pieces of large clusters at a time.

One cluster compact enough to view with large scopes is **M93** (NGC 2447) in Puppis. To find M93, aim your finder scope at the bright double star Xi (ξ) Puppis and then move the telescope 2° northwest. You'll see a cluster much smaller than M41 but very bright — this object really stands out. Changing to a medium-power eyepiece will help to reveal M93's individual stars and will provide a slightly darker background sky to see stars against. (This is a useful technique for looking at small- and medium-size clusters.) Use a magnification that shows M93 about as spread out as your low-power view of M41.

NGC 2244
Mag. 4.8 24' by Chris Floyd

NGC 2158
17.5-inch f/4.5 Reflector 142x
by David J. Eicher

AS LARGE AS THE MOON, Gemini's M35 is visible to the naked eye on dark nights. M35's companion cluster is NGC 2158, a small collection of faint stars 30' southwest of M35's center. Twenty degrees southeast of M35 in Monoceros lies NGC 2244, a bright cluster encircled by the Rosette Nebula.

M35
Mag. 5.1 28' by Mace Hooley

NGC 2158
Mag. 8.6 5' by Lee C. Coombs

M93's stars are slightly fainter than those in M41, and the dimmest members sometimes appear to get lost in the rich Milky Way starfield that surrounds M93. Unlike the stars of M41, M93's members more or less clump together in a band aligned east-west, with a wedge-shaped appendage extending to the north. This object strikingly resembles an arrowhead, something you'll notice right away with a 6-inch or larger scope.

Some 8° north of M93 lies a delightful tangle of 5th- and 6th-magnitude stars known as **M47** (NGC 2422). This cluster is dramatically different than M41. M47 shows a vast range in brightness between its most and least luminous stars — the bright blue stars shine nearly ten magnitudes brighter than the faintest, which are invisible in small scopes. The bright blue stars in M47 are scattered randomly across an even, circular background of dim stars. Because it contains many faint stars, M47 benefits greatly from viewing under a dark, country sky. Under these conditions the lower the power you use, the better M47 looks.

Moving east from M47, you'll encounter **M46** (NGC 2437) some 90' away. This cluster is almost the same size as M41, but its stars are much fainter. If you're using a small telescope in a light-polluted area, you may have trouble spotting M46, but a 6-inch scope from a dark-sky site shows it well. You'll see that nearly all of M46's stars glow at about the same magnitude. Low powers are required to see all of this group's large, circular outline.

But before you pass on to another cluster, switch to moderate or high powers, which will reveal a curious ring-shaped nebula on the cluster's northern edge. This is NGC 2438, a slightly fainter version of Lyra's famous Ring Nebula. NGC 2438 is a planetary nebula — the gaseous remains of a dead star — merely masquerading as a member of M46. NGC 2438 is actually a foreground object. On nights when the seeing is good, try to spot two stars *inside* NGC 2438. These stars are not associated with the nebula, they only appear to be.

Our next object lies northeast of Orion's bright star Betelgeuse and is buried in the middle of the winter Milky Way. **NGC 2244** is a sparkling cluster that energizes a large, glowing wreath of light called the Rosette Nebula. NGC 2244 is quite similar to M41 in that many of its stars are bright, blue-white beacons that can be seen on prac-

M36, M37, and M38
Mags. 6.0, 5.6, and 6.4   12', 24', and 21'   by D. B. Jamison

AURIGA'S TRIO OF BRIGHT CLUSTERS consists of M36 (center), M37 (lower left), and M38 (upper right). M36 and M38 are large star groups visible in finder scopes; M37 is similar to Scutum's rich Wild Duck Cluster.

tically any clear night. They display a wide range of brightnesses, some glowing as dimly as 14th or 15th magnitude. In this respect it is also similar to M41. However, NGC 2244's brightest stars form a box shape rather than a clumpy distribution, and here the similarity ends.

The stars of NGC 2244 are visible in a 3-inch telescope even when the atmosphere is not perfectly steady. A large field of view is required to see the entire cluster because it spans over 1°. At a dark sky site a 6-inch or larger scope reveals the faint glow of the Rosette Nebula encircling NGC 2244.

North of the Rosette Nebula lies Gemini, home of one of the finest clusters in the sky. **M35** (NGC 2168) is an enormous (over 1°-diameter) cluster visible to the naked eye in moderately dark skies. The stars in M35 scatter so widely that your lowest-power eyepiece will provide the best views. The wide range of brightnesses of stars in M35 — from 5th magnitude to beyond visibility — and easy geometric patterns of stars mean this cluster is a good place to start sketching. You can begin with the bright stars and fill in progressively fainter ones. You can even continue your sketch a few nights after you begin. The prominent patterns in M35 — triangles, squares, and arcs — make sketching the stars easy and fun.

If you observe M35 with an 6-inch or larger telescope, you may see a small, concentrated patch of fuzzy light about one Moon-diameter southwest of M35. This is **NGC 2158**, an open cluster similar to M35 but six times more distant. Because of this, its faint stars and arrowhead shape require high power to see clearly.

Auriga's trio of Messier clusters — M36, M37, and M38 — differ from one another and from M41 in substantial ways. **M38** (NGC 1912) is similar to M35 in Gemini because of its great, scattered collection of stars and geometric star patterns. It also has a small companion cluster, NGC 1907, that is visible in large scopes much as NGC 2158 is. M38 contains a wide range of bright and faint stars, in contrast with **M36** (NGC 1960), which leaps out of the background with a dozen bright stars. One of the southernmost stars is a bright pair. While there are bright stars near the center of M36, there is no great central concentration like the one in M41. This is a nice feature of open cluster observing: each cluster has its own individual characteristics.

M37 (NGC 2099) is the "Grand Cluster of Auriga," a wintertime version of the rich and beautiful Wild Duck Cluster in Scutum. The lone bright star in M37, a 9th-magnitude orange star, is the cluster's most distinctive member. M37 is a breathtaking object under a dark sky, when its triangular mass of faint stars includes dozens that spill outside of a low-power field, which leaves dark

NGC 457
Mag. 6.4   13'   by Lee C. Coombs

NORTHERN MILKY WAY CLUSTERS NGC 457 in Cassiopeia and the Double Cluster in Perseus are bright enough to observe with any telescope.

NGC 869 (bottom) and NGC 884
Mags. 4.3 and 4.4   30' and 30'   by Jean Dragesco

Winter's Best Star Clusters

Object	R.A. (2000.0)	Dec.	Magnitude	Size	Tirion	Uranometria
NGC 457	1h19.1m	+58°20'	6.4	13'	1	I: 36
NGC 869	2h19.4m	+57°09'	4.3	30'	1	I: 37
NGC 884	2h22.4m	+57°07'	4.4	30'	1	I: 37
NGC 1907	5h28.0m	+35°19'	8.2	7'	5	I: 97
M38 (NGC 1912)	5h28.7m	+35°50'	6.4	21'	5	I: 97
M36 (NGC 1960)	5h36.1m	+34°08'	6.0	12'	5	I: 97
M37 (NGC 2099)	5h52.4m	+32°33'	5.6	24'	5	I: 98
NGC 2158	6h07.5m	+24°06'	8.6	5'	5	I: 135, 136
M35 (NGC 2168)	6h08.9m	+24°20'	5.1	28'	5	I: 136, 136
NGC 2244	6h32.4m	+04°52'	4.8	24'	11	I: 227
M41 (NGC 2287)	6h47.0m	−20°44'	4.5	38'	19	—
M47 (NGC 2422)	7h36.6m	−14°30'	4.4	30'	12	—
M46 (NGC 2437)	7h41.8m	−14°49'	6.1	27'	12	—
NGC 2438	7h41.8m	−14°44'	10.1	66"	12	—
M93 (NGC 2447)	7h44.6m	−23°52'	6.2	22'	19	—

Object: Designation in Messier or NGC catalog. **R.A. and Dec.:** Right ascension and declination in equinox 2000.0. **Magnitude:** V magnitude of cluster. **Size:** Diameter in arcminutes or arcseconds. **Tirion:** Chart on which object appears in *Sky Atlas 2000.0* by Wil Tirion. **Uranometria:** Volume number and chart on which object appears in *Uranometria 2000.0* by Wil Tirion, Barry Rappaport, and George Lovi.

lanes and star chains meandering through the eyepiece.

NGC 457 is the finest cluster in Cassiopeia. Large binoculars or small scopes show the two bright "eyes" in this star group nicknamed the Owl Cluster. The brightest of the eyes, orangy Phi (φ) Cassiopeiae, is one of the brightest stars known. The owl's body is composed of doublets and triplets of mainly yellow stars, and two thin wings stretch out from the body. The eastern wing contains an exceptionally red star about halfway along its length. High powers and averted vision reveal the glitter of background stardust, the light from dozens of faint, unresolved stars.

King of the northern sky's star clusters is the **Double Cluster** in Perseus, NGC 869 and NGC 884. Located just south of the bright "W" of Cassiopeia, the Double Cluster is easily visible to the naked eye as an oval glow even on nights of average transparency. The easternmost component, NGC 869, is the brighter of the two clusters. Use the lowest possible magnification to capture the two clusters in a single field of view. The central luminary in this group is located at what appears to be the neck of a humanlike stick figure of stars. To the south is a little circlet that forms the head, while to the north are two straight legs. High power reveals an extraordinary number of stars in this area, including some faint red stars.

NGC 884, the western half of the Double Cluster, is equal in size to its neighbor and only slightly fainter. Notice the variety of subtle star colors in NGC 884: blue, blue-white, white, yellow, orange, and even red stars are visible. Seldom does a group of stars allow you to see virtually every color of star in one small area.

While you warm up indoors after viewing the bright star clusters of winter, think about what you've seen.

Quite different from some of the summer clusters, the winter clusters are quite young. M41, a moderately young cluster, is approximately 190 million years old. Others range from the exceedingly young — like the Double Cluster's two members (3.2 million and 5.6 million years) and NGC 2244 (3 million years) — to the aged cluster NGC 2158 (3.2 billion years). Its great age reveals that NGC 2158 is not a part of the Orion arm — it is a distant background object lying 15,000 light-years away compared with M41's distance of 2,000 light-years. NGC 2158 is visible *through* the Orion arm of the Galaxy.

The winter clusters are typical in size, ranging from 12 light-years in diameter (M47) to 60 light-years across (each component of the Double Cluster). Most of the bright clusters, including M41, are 20 to 25 light-years in diameter. Starlight, traveling at 186,000 miles per second, takes twenty-five years to traverse the diameter of a cluster like M41. Standing under the crisp skies of winter, staring into a chilled eyepiece, you'll get the message that hits hardest — the universe is *big*.

Armed with a telescope and wrapped up to keep warm, you can witness a great deal of the Galaxy by observing bright winter clusters. On one level you can see the diversity that characterizes open star clusters — the bright central stars of M41 and M37, subtle star colors in the Double Cluster, and the fuzzy splash of light called NGC 2158. After observing these star groups, you'll see that each open cluster has qualities that separate it from the rest. On another level you can see that in the sky's great sea of black a few small clumps of stars mark an ancient, fundamental process in the Galaxy — the formation of new stars from old ones. □

Dramatically Diverse Globulars

Even in a small telescope you'll find that globular clusters come in a variety of sizes, densities, and brightnesses.
by Chris Schur

"Globular clusters are all alike — every single one is nothing more than a dim, fuzzy ball of light!" I can't keep count of how many times I've heard someone say that at a star party. Yet it couldn't be farther from the truth. Globular star clusters offer great variety for amateur astronomers. Anyone with a small telescope and a sharp, patient eye can see rich details in just several nights of globular hunting.

Globular clusters are unique, ancient pieces of the Galaxy. Massive spheres of stars, they reside in the Milky Way's halo, far from the galactic disk. Astronomers believe globulars formed at the same time as the rest of the Galaxy but for unknown reasons didn't succumb to the gravitational tug of the developing galactic disk. Instead, they remained independent of the Galaxy, orbiting it like bees around a hive. Their orbits periodically carry them through the galactic plane, sweeping them free of gas and dust necessary for star birth, sterilizing them. Therefore, globular clusters contain old, cool, slow-burning stars that softly glow in shades of yellow and orange. These stars are pristine links to the Galaxy's ancient past.

One of the best places to see these stars is in **M13** (NGC 6205), the Hercules Cluster. Although several globulars are larger and brighter than M13, the Hercules Cluster is unrivaled for viewers in the north because it rises high in the sky. Located at +36° declination, glowing at magnitude 5.9, and measuring 16.6' across, M13 is barely visible to the naked eye on dark nights. It appears like a 6th-magnitude star on the western side of Hercules' distinctive "Keystone" asterism.

To find M13, aim your telescope at the 4th-magnitude double star Eta (η) Herculis, the northwest corner of the Keystone. Move the scope 2.5° to the south and you will see M13 as a small, fuzzy "star" with a brighter center. The cluster forms a compact isosceles triangle (20' on a side) with two 7th-magnitude stars, one positioned south of the cluster and the other to the east. Several dozen fainter stars lie scattered uniformly within 2° of the cluster.

The smallest telescopes and binoculars portray M13 as a tiny round disk devoid of stars. Its gray-green glow betrays a star but fails to display cluster stars, which gives the impression of a washed-out planet. However, the cluster's appearance dramatically changes given a few more inches of aperture.

On nights when the atmosphere is steady, a 4-inch telescope at moderate powers shows a few of the bright-

M56
Magnitude 8.3 Size 7.1' by Lee C. Coombs

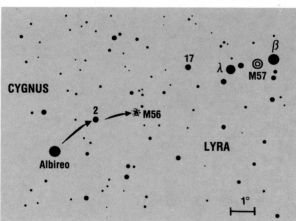

TINY, CONDENSED M56 lies in the constellation Lyra in a rich field of stars. To find it, star hop from Albireo to 2 Cygni and on to the cluster.

est stars near the cluster's edges, red giants that glow feebly at 11th magnitude. The overall view is that of a glowing ball of light with an intense center and a mottled appearance at the edges. During moments of good seeing the atmosphere permits light from the brightest stars to pass and you can see these stars as distinct points of light. A 4-inch scope is generally the smallest that reliably shows M13 as a cluster of stars, not just a nebulous glow.

Larger apertures provide a stunning view of the Her-

M13
Magnitude 5.9   Size 16.6′   by Jack Newton

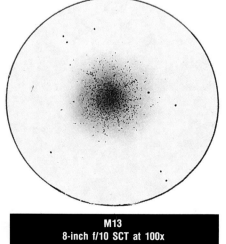

M13
8-inch f/10 SCT at 100x

THE GREATEST GLOBULAR of the northern sky, M13 in Hercules is easily resolved in small telescopes; large scopes show lanes of stars winding away from its center. It lies on the western side of Hercules' Keystone asterism.

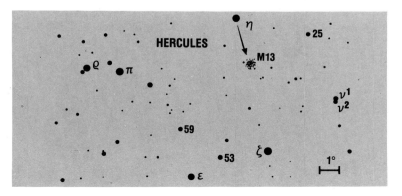

M22
Magnitude 5.1   Size 24.0′   by Scott Hicks

M55
Magnitude 7.0   by Martin Germano

M75
Magnitude 8.6   by Jack Newton

RICH, LOOSE M22 is the finest cluster in Sagittarius. Individual stars in M22 are visible even with a 4-inch telescope. Even less condensed is M55, visible as a faint cloud of light in small scopes. In marked contrast, nearby M75 is a bright, condensed ball of light that is difficult to resolve into stars.

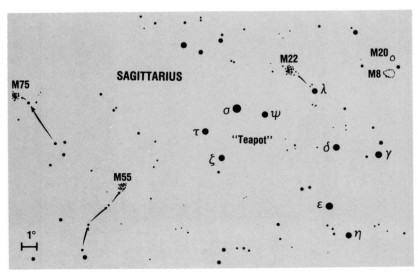

M107
Magnitude 8.1 Size 10.0' by Jack Newton

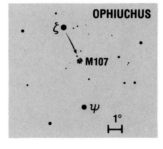

A TINY KNOT of stars, M107 in Ophiuchus can be found by star hopping from Zeta Ophiuchi south to a triangle of three dim stars. The cluster lies immediately to the south.

cules Cluster. An 8-inch scope at 100x shows a field gleaming with stars: the big, hazy ball of light appears almost three-dimensional because of a spray of tiny stars that covers it.

In a 16-inch or 18-inch telescope the density of resolved stars grows from a few dozen encircling the cluster's edges to layers of hundreds of star points converging on the cluster's bright, central, yellow glow. At high powers the stars are too numerous to count, and undoubtedly many of the faint stars spilling out of a single field, scattered several arcminutes in every direction, belong to the cluster. The view is one that a photograph simply can't duplicate.

A few tricks of the trade can help you see more when you observe M13 and other globular clusters. Generally, high powers help to resolve the tiny, faint stars in globulars. However, don't use a magnification that is too high, or you'll simply get mushy, bloated star images. A crisp, low-power view beats a soft, high-power view. Try using 100x to 150x, and if you can't resolve the cluster, try again later when the atmosphere may be steadier. Magnifications much in excess of 200x normally only magnify atmospheric turbulence.

A combination of averted vision — glancing toward the side of the telescope's field — and very slowly sweeping the telescope side to side helps reveal faint details in clusters. This technique works because the eye's faint light receptors are acutely sensitive to motion. For example, this technique shows an extremely bright, condensed, ball-like center in the Hercules Cluster. Used with a large telescope, sweeping and averted vision reveal apparent dark "voids" and long streamers of stars that wind away from the center of M13 like spidery legs.

Field sweeping also helps in spotting **M56** (NGC 6779), located east of Hercules in the small constellation Lyra. Unlike M13, M56 lies in a rich starfield and is a small, condensed globular that is difficult to resolve because its stars are faint and tightly packed together. To locate M56, aim your finder scope midway between Gamma (γ) Lyrae and Albireo, the star that marks the base of the cross in Cygnus. There you'll see a close pair of 6th- and 9th-magnitude stars. Visible in finders as a small patch of nebulous light, M56 lies 1° southeast of this pair.

The stars in M56 together glow at magnitude 8.3 and are tightly packed into a diameter of only 7.1' — less than half the size of M13. By contrast, M56 is much harder to resolve into stars. Once you've located the cluster, switch to moderately high power (15x per inch of telescope aperture) and slowly sweep back and forth over the cluster. With a 6-inch telescope you'll see a disk of gray-green light encircling a slightly brighter center, set in a field splattered with dozens of faint stars. The view of M56 changes little with increased aperture. The largest backyard telescopes, those in the 16-inch or larger range, resolve some of the cluster's stars on nights of steady seeing.

Follow the Milky Way south to find a dramatic contrast to M56, Sagitta's **M71** (NGC 6838). This cluster is one of the richest globulars visible in small telescopes and is partly resolvable in a 6-inch telescope. In fact, its stars are so loosely assembled that for years some astronomers proposed that M71 may be an exceedingly rich open star cluster.

To find M71, start from Zeta (ζ) Sagittae, a bright star in the center of the constellation's distinctive arrow-shaped asterism. Star hop 1.5° east-southeast and you'll arrive at a rich field containing a 6th-magnitude double star and M71. (The loose open cluster Harvard 20 lies just south of this field.)

Telescopically M71 is a delight. The cluster's magnitude and size are virtually identical to those of M56, but M71 offers a far better view. A 4-inch scope at high power resolves a few of the group's brightest stars around the edges. Six-inch telescopes reveal that M71 has a slightly noncircular shape and a dramatically bright center. In large scopes M71 really shows its stuff: a 12-inch scope resolves the cluster clear across its face, unveiling numerous white and pale yellow-orange pinpoints apparently hanging in front of a hazy gray-green backdrop. Although it isn't as large or bright as M13, M71 is nonetheless one of the finest globular clusters for backyard telescopes.

Three globulars are tucked in the constellation Ophiuchus. Similar in size and brightness and separated by a mere 3°, **M10** (NGC 6254) and **M12** (NGC 6218) are easily visible in small scopes. Each glows at magnitude 6.6 and covers about 15' of sky. To find them, start by placing your finder scope on the 3rd-magnitude star Zeta (ζ) Ophiuchi and move 8° northeast to 5th-

M15
Magnitude 6.4   Size 12.3′   by Jack Newton

M15 IN PEGASUS is the finest globular cluster in the autumn sky. It is visible in the spring after Pegasus rises early in the morning. Look for M15 3.5° northwest of Enif, one of the constellation's brightest stars.

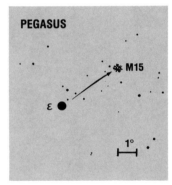

M71
Magnitude 8.3   Size 7.2′   by Martin C. Germano

Globulars for Small Scopes

Object	R.A. (2000.0)	Dec.	Mag.	Size	Tirion	Uranometria
M107	16h32.5m	−13°03′	8.1	10.0′	15	II: 291
M13	16h41.7m	+36°28′	5.9	16.6′	8	I: 114
M12	16h47.2m	−1°57′	6.6	14.5′	15	II: 247
M10	16h57.1m	−4°06′	6.6	15.1′	15	II: 247
M22	18h36.4m	−23°54′	5.1	24.0′	22	II: 340
M56	19h16.6m	+30°11′	8.3	7.1′	8	I: 118
M55	19h40.0m	−30°58′	7.0	19.0′	22, 23	II: 379, 380
M71	19h53.8m	+18°47′	8.3	7.2′	8, 16	I: 162
M75	20h06.1m	−21°55′	8.6	6.0′	22, 23	II: 343
M15	21h30.0m	+12°10′	6.4	12.3′	16, 17	I: 210

Object: Designation in Messier catalog. **R.A. and Dec.:** Right ascension and declination in equinox 2000.0. **Mag.:** V magnitude of cluster. **Size:** Diameter in arcminutes. **Tirion:** Chart on which object appears in *Sky Atlas 2000.0* by Wil Tirion. **Uranometria:** Chart on which object appears in *Uranometria 2000.0* by Wil Tirion, Barry Rappaport, and George Lovi.

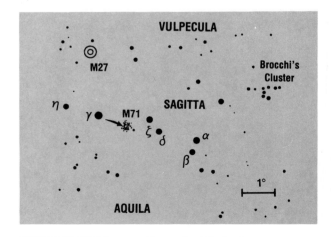

GLOBULAR M71 is so easily resolved that astronomers once thought it may be an open cluster. Locate M71 near a bright double star at the center of Sagitta.

M10
Magnitude 6.6 Size 15.1' by Bill Iburg

M12
Magnitude 6.6 Size 14.5' by Bill Iburg

PAIRED IN OPHIUCHUS, globular clusters M10 and M12 are nearly identical except for star densities. M10 has a bright, condensed core; M12 is easily resolved into myriad stars.

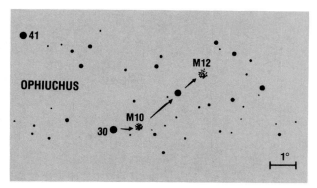

magnitude 30 Ophiuchi. M10 lies 1° west of this star and M12 3° northwest of M10. Both clusters appear like smaller and dimmer versions of the Hercules Cluster. However, M10 has a tightly packed distribution of stars, whereas M12 is somewhat looser. High magnifications reveal star chains and voids across parts of this cluster.

Now return to Zeta Ophiuchi and move 3° south-southeast to find **M107** (NGC 6171). An unusual globular cluster, M107 glows at magnitude 8.1 and covers 10' of sky, yet it is deceptively hard to resolve. A 10-inch scope does the job under good seeing, but a 12-incher provides a more satisfying view. Large telescopes show that M107's core appears like a grainy, glistening clump.

Sagittarius abounds in globular clusters, as it does with other types of deep-sky objects. Three of the constellation's best globulars are **M22** (NGC 6656), **M55** (NGC 6809), and **M75** (NGC 6864). M22, located 3° northeast of the bright star Lambda (λ) Sagittarii, is in fact one of the finest globulars in the sky. Shining at magnitude 5.1 and measuring 24.0' across, M22 exceeds the Hercules Cluster in every aspect. It is also far easier to resolve; a 4-inch scope shows moderate resolution. However, its southern declination hurts these impressive statistics for viewers in the Northern Hemisphere.

M55 and M75 lie in the eastern corridor of Sagittarius, away from the Teapot asterism. Locate M55 by sweeping 7° east and 1° south of Zeta Sagittarii. M55 is a large, loose globular cluster that is easily resolved at moderate powers. A 6-inch scope shows it as a 7th-magnitude glow peppered with faint stars. M75, which lies near the Sagittarius-Capricornus border, is a 8.6-magnitude cluster with stars compressed into a 6' diameter. Resolving this tight cluster constitutes a fine challenge for large telescope owners.

The grand globular of autumn, Pegasus' **M15** (NGC 7078), is a dazzling sight that resembles a scaled-down version of the Hercules Cluster. M15 lies immediately west of a solitary 7th-magnitude star some 3.5° northwest of Enif (Epsilon [ε] Pegasi). With a good finder scope it's a snap to find because M15 is clearly a fuzzy ball of light even in a 6x30mm finder. Once you've located it, increase your scope's magnification to 15x per inch of aperture and carefully sweep over the field. M15 lies in a rather sparse field of faint stars but is itself a beautiful sight.

A 4-inch scope begins to resolve the cluster's edges into stars. A 6-inch scope shows many stellar dots sprinkled across M15 and an intense glow in the cluster's center. The high-power view with a large backyard telescope is stunning: star chains spill out of the field in all directions. Gazing at M15, I get the impression I'm watching myriad other worlds and wonder if perhaps something is looking back at me.

Because these clusters are suspended far away in the Milky Way's halo, viewing them requires peering through slices of the Galaxy itself. Observing some globulars — M10, M12, M22, M55, and M75 — from our position in the Galaxy makes us gaze through a thick section of the Galaxy's disk. Consequently we see these clusters in rich starfields. However, clusters that lie away from the center of the Galaxy — M13, M15, and M56 — lie in relatively sparse fields of stars because our vantage point makes us look through thin parts of the Galaxy's disk. However, whether they lie among numerous, sparkling stars or in just a handful of dim points of light, globular clusters offer pretty views for users of small telescopes. □

These objects are best visible in the spring evening sky.

Scanning the Scutum Starcloud

Rich in dazzling star clusters and challenging nebulae, the Scutum starcloud offers many delights for summertime observers.
by Phil Harrington

"It is absolutely the gem of the Milky Way," said the great observer Edward Emerson Barnard, referring to his favorite part of the Milky Way, the magnificently rich Scutum starcloud. Following the star-studded plane of the Galaxy southward from Cygnus through Aquila and on toward Sagittarius, the Scutum cloud is an immediate standout. The constellation itself is not particularly bright or large, but the milky glow centered on the four stars that make up the Shield is eye catching under a dark sky.

The easy visibility of the Scutum starcloud is caused by a contrast effect. The starcloud is simply a rich region of unresolved stars — a prominent portion of the Milky Way cut off by dark bands of dust.

If Scutum is the gem of the Milky Way, then **M11** (NGC 6705) is Scutum's crown jewel. Called the Wild Duck cluster because its richness and triangular shape resembled a flock of ducks in flight to early observers, M11 is a magnificent sight through binoculars and telescopes and is one of the prettiest open clusters in the sky.

M11 was first sighted by Gottfried Kirch of the Berlin Observatory in 1681. Kirch described the cluster as a "small, obscure spot with a star shining through." Eighty-three years later French comet hunter Charles Messier saw the object and called it "a cluster of a great number of small stars. In a three-foot [focal length] instrument it looks like a comet." William Herschel thought the stars of M11 appeared "divided into five or six groups," an opinion echoed by D'Arrest.

Any search for M11 should start at the star Lambda (λ) Aquilae. While viewing through a finder scope or binoculars, curve to the southwest past 4th-magnitude Iota (ι) Aquilae to Eta Scuti. Continuing westward you will arrive at a crooked rectangle of 6th- and 7th-magnitude stars. M11 lies less than 1° southeast of this rectangle. Because it shines at magnitude 5.8, M11 will be immediately visible in your finder scope. Once you've located the cluster you may even be able to make it out with the naked eye under a perfectly black sky.

Through finder scopes and low-power binoculars M11 reveals a bright, unresolved glow highlighted by an 8th-magnitude orange star near its center. A 6-inch telescope at 50x shows M11 as a collection of dozens of glowing pinpoints spread across 14'. Except for the lone central star, the members of M11 glow at 11th-magnitude or fainter.

At first glance M11 may appear so rich that it looks like a globular cluster. A closer look reveals M11 is lopsided. The cluster's brightest stars are shaped in a distinctive arrowhead pattern that eighteenth-century English observer William Smyth noted in his classic catalog of deep-sky objects. Smyth wrote that M11 resembles "a flight of wild ducks."

Large telescopes and high powers resolve the core of M11. A 13-inch reflector at 100x shows a breathtaking view of a field strewn with hundreds of stars, each slightly bluish white.

Of the many dark nebulae that surround M11, **B111** and **B119a** are the most obvious. They are visible north of M11 as large, kidney-shaped patches almost touching each other at their southern tips. Although visually these nebulae appear to be nearly identical in size, photographs show that B111 is actually twice as large as B119a. Adding to the area's attractiveness is a "peninsula" of half a dozen 8th- to 10th-magnitude stars wedged between the nebulae.

After you've observed the two dark nebulae, move back to the northwesternmost star in the small rectangle

M11
Magnitude 5.8 Size 14' by George R. Viscome

The Scutum Starcloud
by Ronald Royer

NGC 6712 and IC 1295
Magnitudes 8.2 and 15.0   Sizes 7.2' and 86"   by Martin C. Germano

M26
Magnitude 8.0   Size 15'   by Lee C. Coombs

near M11. This is **R Scuti**, a bright example of the RV Tauri class of pulsating variable stars. RV Tauri stars are famous for vacillating irregularly between shallow and deep brightness minima. The difference between two succeeding dips in brightness may be as great as three or four magnitudes.

In the case of R Scuti, the star's brightness typically alternates between 5th and 6th magnitude over 140 days. At the same time its spectral classification fluctuates between type-G5 and type-K. A secondary period causes R Scuti to drop to about magnitude 8.2 every fourth or fifth cycle. This complex behavior is thought to be caused by differing rates of expansion and contraction by each of the star's many internal layers.

Open cluster **NGC 6704** has always struck me as a sort of poor man's M11. Located 1° north of its impressive neighbor, NGC 6704 is visible in small telescopes only as a hazy smudge of light. A 6-inch scope is the smallest aperture that resolves the brighter stars in this tightly packed open cluster. Viewed through large backyard telescopes, NGC 6704 shows numerous stars sprinkled across a 6'-diameter.

Now move 4.5° southwest of M11 to the 4th-magnitude star Alpha (α) Scuti. Alpha is an excellent jumping-off point for starhoppers searching for nearby treasures. If you switch to your widest field eyepiece and center your attention on Alpha, our next target might just make it into the view.

NGC 6664 is a pretty open cluster found less than a Moon's diameter east of Alpha Scuti. Binoculars and small telescopes reveal a soft, irregular blur highlighted by a single 10th-magnitude star. Six-inch and larger instruments begin to reveal many more stars, most forming a pattern resembling the letter "U." NGC 6664 is thought to consist of approximately fifty luminous type-B stars.

Extending for over half a degree around Struve 2325 is the faint reflection nebula **IC 1287**. To see this object you must use superlative optics on a very transparent night. I have glimpsed this nebula on rare occasions through my 13-inch reflector as a homogeneous, oval glow.

NGC 6649 resides 45' northeast of IC 1287. A little-observed open cluster, NGC 6649 is populated by some fifty stars spread across 6'. None of these stars shines brighter than magnitude 11.6, although a 9th-magnitude field star is superimposed on the cluster's

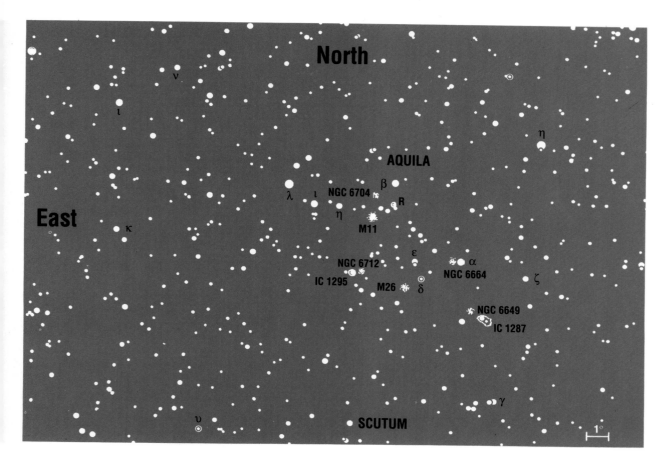

Deep-Sky Objects in Scutum

Object	R.A. (2000.0)	Dec.	Mag.	Size
IC 1287	18h31.3m	−10°50'	—	44' by 34'
NGC 6649	18h33.5m	−10°24'	8.9	6'
NGC 6664	18h36.7m	−08°13'	7.8	16'
M26	18h45.2m	−09°24'	8.0	15'
R Scuti	18h47.5m	−05°42'	4.5-8.2	140d
NGC 6704	18h50.9m	−05°12'	9.2	6'
B111/119a	18h51:	−5°:	—	120' by 120'
M11	18h51.1m	−06°16'·	5.8	14'
NGC 6712	18h53.1m	−08°42'	8.2	7.2'
IC 1295	18h54.6m	−08°50'	15.0P	86"

Object: Designation in Messier, NGC, or other catalog. **R.A. and Dec.:** Right ascension and declination in equinox 2000.0. **Mag.:** V magnitude; subscript P denotes photographic magnitude. **Size:** Diameter in arcminutes or arcseconds; period in days for variable stars.

southwestern border. In an 8-inch telescope this cluster appears like a coarse, oval glow that is best seen at lower magnifications.

M26 (NGC 6694) is a rich open cluster populated by thirty stars of magnitude 10.3 and fainter. Most blend into a nebulous glow through small telescopes, with only about four or five points of light clearly seen in 4-inch scopes. Tripling the aperture increases the number of resolved stars to about two dozen spread over a hazy background glow.

Once you have examined M26, pause and let the stars drift across your eyepiece field for seven minutes. Peek through your finder scope and you'll find three 6th-magnitude stars that form a slightly crooked row. Switching to your main instrument, gently nudge the telescope northward about 1° to find **NGC 6712**. It's a wonder that Messier missed this magnitude 8.2 globular cluster in his survey of the area. You might be able to pick up its dim glimmer through 7x50 binoculars on dark nights. The cluster's relatively loose structure permits partial resolution through 4-inch telescopes; larger instruments hint at the irregular structure of the cluster's nucleus.

Shifting about one eyepiece field to the east of NGC 6712, we encounter the planetary nebula **IC 1295**. Although many catalogs list this object as having a very low surface brightness, I found it easily visible in an 8-inch reflector. Moderate powers revealed a slightly oval disk highlighted by a brighter center. Little additional detail is visible through larger instruments, but long-exposure photographs record a 17th-magnitude central star along with the nebula's subtle ring-like structure.

It's easy to see why Barnard called the Scutum starcloud the "gem of the Milky Way." Few portions of the sky cram so many fine and varied deep-sky objects into so small a zone as does the tiny constellation Scutum. The Scutum cloud shines with a luster unequaled all along the northern Milky Way.

These objects are best visible in the summer evening sky.

Splashy Summer Star Clusters

Ready to give up on naked-eye observing? Discover four dazzling open star clusters visible to the eye alone and grand sights in binoculars.
by Alan Goldstein

The summer sky contains many beautiful open clusters. These collections of stars consist of anywhere from several dozen to several thousand members scattered loosely over volumes of space from less than twenty to more than two hundred light-years in diameter. Although they lie at great distances, the largest and brightest open clusters still appear large in our sky because of their great physical sizes. The closest of these open star clusters are bright enough to detect with the eye alone under a dark sky.

Open clusters vary widely in appearance. Some are quite dense, resembling miniature globular clusters. Others are so poor they almost blend in with the background star field. Their current appearances depend largely on their ages: Formed from gas and dust, the newly born stars in clusters slowly move apart and dissipate as they orbit the center of the Galaxy.

One of the brightest summer clusters is **M7** (NGC 6475) located in the tail of Scorpius. Although the brightest members of M7 glow at about 6th magnitude, the integrated magnitude of the entire cluster is 3.3, making it a naked-eye object even under mildly light-polluted skies. Under dark rural skies, M7 and its neighbor M6 look like two bright knots in the southern Milky Way.

In binoculars and small telescopes, the brightest members of M7 resemble a letter "K" turned on its side. Slightly more than 1° across, M7 is best observed with binoculars or an RFT. In fact, any view less than 2° in diameter diminishes this cluster's spectacular appearance. Because half of its 130 members are brighter than 10th magnitude, M7 is a good subject to impress your non-astronomer neighbors or family members even under suburban skies.

M7 lies in front of a mottled Milky Way star field full of bright patches and ribbon-like dark nebulae. Can you see them? A wide-field telescope or large binoculars should do the trick under very dark skies. A field diameter between 3° and 5° is required. The bright stars in M7 will actually interfere with your ability to pick up the subtleties in background glow. Another challenge is finding the tiny globular cluster NGC 6453, which is located on the western edge of M7 although actually it is considerably farther away from us.

About 3.5° northwest of M7 lies **M6** (NGC 6405). Visible to the naked eye as a magnitude 4.2 glow, this is another splashy cluster in the summer sky. Many

M7
Mag. 3.3   Size 80'   by Alfred Lilge

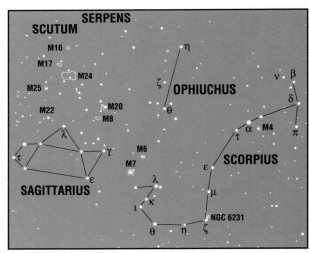

observers see the brighter stars of this 130-member cluster as forming the shape of a butterfly wing. It is therefore not surprising that this cluster is often called the Butterfly Cluster. On nights of mediocre seeing, the butterfly pattern may not be obvious and the brighter stars form a "W"-shaped pattern. Looking closely at the fainter stars at the southern end of the cluster should make the W-pattern more apparent.

The brightest stellar member of M6 is **BM Scorpii**.

Sagittarius-Scorpius region by Chris Floyd

It is an irregular variable, orange in color (spectral type K0 to K3), with a period of roughly 850 days. Catch it at its brightest and it stands out at magnitude 6.0. If you spy it at minimum brightness (8.1 magnitude), BM Sco will still be conspicuous and very red. If you have ever thought about dabbling in variable star observing, following BM Sco's semi-unpredictable pattern of variation makes a good project.

Although **M25** (IC 4725) is a spectacular cluster in Sagittarius, it is not as richly packed with stars as M6 or M7. This cluster contains over fifty members approximately 6th magnitude and fainter. Its total magnitude is 4.6, making it a naked-eye object under a dark sky. M25 is a showpiece for binoculars and rich field telescopes. The 32' diameter cluster accents an already rich star cloud toward the center of our galaxy. The brightest members of this cluster glow with a variety of colors that can be observed in binoculars or a

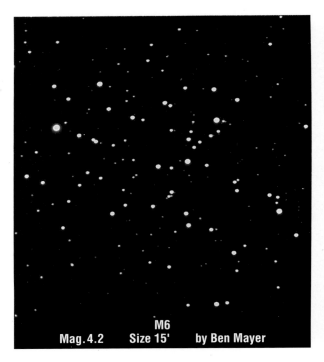

M6
Mag. 4.2   Size 15'   by Ben Mayer

M25
Mag. 4.6   Size 32'   by Lee C. Coombs

NGC 6231
Mag. 2.6   Size 15'   by Mike Mayerchak

Summer Naked-Eye Clusters

Object	R.A. (2000.0) Dec.	Mag.	Size
NGC 6231	16h54.0m −41°48'	2.6	15'
M6	17h40.1m −32°13'	4.2	15'
M7	17h53.9m −34°49'	3.3	80'
M25	18h31.6m −19°15'	4.6	32'

Object: Designation in Messier or NGC catalog. **R.A. and Dec.:** Right ascension and declination in equinox 2000.0. **Mag.:** V magnitude. **Size:** Diameter in arcminutes.

low-power telescope. In larger telescopes, color tends to become muted and more difficult to see because of the intensity of starlight.

NGC 6231 would be one of the all-time favorite open clusters if it was not located so far south. If your latitude lets you observe the southern reaches of Scorpius, this is a "must-see" star cluster. NGC 6231 contains about 120 stars and is very striking because of the preponderance of bright blue-white members. NGC 6231's brilliant members together shine at magnitude 2.6, making it rather impressive even if it lies relatively near the horizon. From anywhere south of +40° latitude this cluster is an easy and spectacular naked-eye object.

Spend a few nights gazing up at these spectacular summer clusters. Whether you observe these star groups with the naked eye, binoculars, or a small telescope, you'll see that they represent some of the prettiest sights in the summer sky. □

These objects are best visible in the summer evening sky.

Southern Clusters of All Ages

The color and density of a cluster can tell you a lot about the age of the cluster's stars.
by Gregg D. Thompson

Scattered along the Milky Way, groups of stars called open clusters offer you a chance to peer into the history of the universe. These clusters contain a record of their past: astronomers study the light from stars in open clusters to discover everything from the ages to the distances and sizes of clusters. Many of the finest open clusters lie in the southern sky, invisible to Northern Hemisphere observers. Nine of the southern sky's best clusters are intrinsically brilliant objects visible to the naked eye.

Of all the bits of knowledge astronomers collect about open clusters, age is the most revealing. Because stars change as they mature, the brightnesses and colors of open cluster stars provide clues to the cluster's age. In general, the bluer the stars, the younger the cluster; the redder the stars, the older the cluster. This in turn lets astronomers uncover the cluster's past and look into its future.

Open clusters are typically little groups of a few dozen to a few hundred stars that form from gas clouds. Open clusters live their lives in the Milky Way's disk, slowly rotating about the galactic center. They are perishable: few open clusters last longer than one or two rotations around the center of the Galaxy. Gravitational tugs from passing clouds of dust and other objects eventually tear away the stars of most open clusters and disperse the stars throughout the Galaxy's disk.

The southern sky's brightest clusters vary greatly in age and characteristics. The youngest is NGC 4755, a group of stars in Crux, one third the size of the Moon. Known as the Jewel Box because of its sparkling blue-white and orange stars, NGC 4755 formed about 7

NGC 4755   Crux

NGC 3293   Carina

NGC 3766   Centaurus

IC 2602   Carina

IC 2391   Vela

NGC 6067   Norma

NGC 2516   Carina

million years ago — very recently in astronomical terms. The cluster is dominated by young, extremely hot, blue and white stars. These stars are intrinsically bright and make the Jewel Box faintly visible to the naked eye even though the cluster lies at the great distance of 7,500 light-years.

The Milky Way clouds in Carina hold another young cluster. A naked-eye object 8,500 light-years away, NGC 3293 is similar in age and appearance to the Jewel Box. This cluster lies in a busy area of the Galaxy within the so-called Carina OB1 and Nebular complex, a region of dust, gas, and young stars. If you aim a pair of binoculars at NGC 3293, you'll see that NGC 3293 is a rich cluster. And if you aim a medium-sized telescope at NGC 3293, you'll see over two hundred stars within the cluster's boundary.

Lying near the famous Eta Carinae Nebula, NGC 3766 is intrinsically one of the most luminous clusters in the Milky Way. It is also young, estimated to be about 20 million years old. In our sky NGC 3766 is a faint naked-eye object because it lies at the great distance of 6,200 light-years. Many of the brightest stars in NGC 3766 are blue and white, but there are also yellowish stars and three red stars. The yellow and red stars are relatively cool stars visible because they are extremely brilliant.

Astronomers estimate that the cluster IC 2602 in Carina is between 4 million and 20 million years old. Sometimes called the "Southern Pleiades," this cluster is bright and splashy even when viewed with the naked eye. IC 2602 glows with the light of a 2nd-magnitude star and covers an area larger than that of the Full Moon. The most brilliant stars in IC 2602 are supergiant stars — extremely large and luminous objects — that shine blue and blue-white.

Young groups of stars also lie in Vela and Norma. IC 2391 in Vela (30 million years old) shines at magnitude 2.5 and has a diameter larger than that of the Moon. It contains thirty stars. NGC 6067 is a rich cluster occupying the central position in the Norma star cloud. Although NGC 6067 lies 6,800 light-years away, it is a naked-eye object in our sky. In binoculars, look for nine orange-colored giant stars among the mostly blue and white crowd. Although some colorful stars are visible in this cluster,

NGC 3114 Carina

NGC 3532 Carina

NGC 6067 is still relatively young at 78 million years and is dominated by blue and white stars.

Carina's NGC 2516 and NGC 3114 are middle-aged open clusters that have a richer variety of star types than younger clusters have. NGC 2516, a group shining at magnitude 3.8 and measuring 30' across, is one of the most colorful clusters in the southern sky. This group is about 110 million years old and contains three red stars and a smattering of yellow and orange stars. As old as NGC 2516, NGC 3114 is a rich cluster that lies about twice as far away as NGC 2516. Compare the two in binoculars and you'll see a great difference in appearance. Although the two contain similar types of stars, the difference in distance makes NGC 3114 appear much smaller and its stars less colorful.

Also located in Carina, NGC 3532 is much older than the other bright clusters in the southern sky. At 200 million years old, this group of stars has a substantial collection of highly evolved stars with abundances of heavy metals. Observe this group with binoculars and you'll see a preponderance of yellow and orange stars but not many blue stars. The stars in this group are much closer to the ends of their lives than those in clusters like the Jewel Box.

NGC 3532 is big and split into two distinct stellar lanes that together contain one hundred stars brighter than 14th magnitude. NGC 3532 has almost completed one revolution about the center of the Galaxy. In another 300 million or so years this cluster will have orbited the Galaxy twice, and by then the cluster may be torn apart and its stars distributed throughout the Galaxy's disk.

Young or old, open star clusters reveal a piece of the universe's past. Looking at them allows us to see a "still frame" from the great movie that is the changing universe. If you live in the Southern Hemisphere, compare these bright clusters and see what they tell you. If you are a northerner, a trip down south will show you some of the most magnificent objects in the heavens. □

To see these objects you must observe from a southern latitude.

The Jewel Box Cluster
National Optical Astronomy Observatories photo

M41
Mag. 4.5   Size 38'   by George R. Viscome

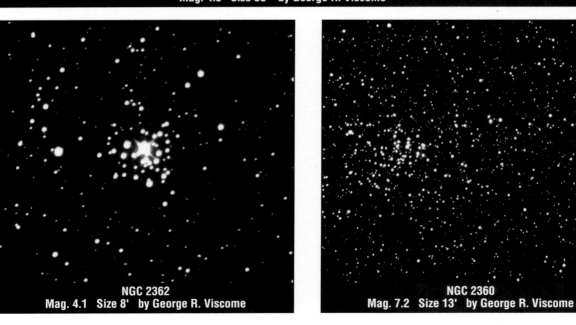

NGC 2362
Mag. 4.1   Size 8'   by George R. Viscome

NGC 2360
Mag. 7.2   Size 13'   by George R. Viscome

Exploring Open Clusters in Canis Major

Filled with sparkling star groups, Canis Major is a powerful attraction on chilly winter nights.

Phil Harrington

Stretching from horizon to horizon and passing through the zenith, the Winter Milky Way is a glorious haven for deep-sky observers. As you scan along the hazy glow of our Galaxy, you'll occasionally see islands of stars — open clusters. These groups of young stars formed from clouds of interstellar gas and dust, and will ultimately break apart as they revolve about the center of the Galaxy. Some of the finest examples of open clusters on the edge of the Milky Way are within the constellation of Canis Major, the Large Dog.

Canis Major is easily found southeast of Orion thanks to the presence of its brightest star, **Sirius** (Alpha Canis Majoris). At magnitude –1.4, Sirius is the brightest star in the sky. It shines like a brilliant blue-white diamond suspended against a velvety black field of stardust. Nearly 4° south of Sirius is the pre-eminent open cluster in Canis Major — **M41** (NGC 2287). M41 is bright enough to be visible to the naked eye on dark nights and was mentioned as a curious "cloudy spot" by Aristotle, who noted it in the year 325 B.C. From a dark, rural site, you can easily duplicate Aristotle's observation and use the naked-eye visibility of M41 to gauge the sky's transparency.

Today we know that M41 consists of about 80 stars, most of which glow in subtle hues of yellow, white, or blue-white. M41's brightest star is slightly brighter than 7th magnitude; 20 of its stars are brighter than 10th magnitude and are therefore visible in large binoculars and finderscopes. A 3- to 4-inch telescope resolves about half of the stars in M41, while an 8-inch instrument shows the cluster in all its glory. Most observers immediately notice that four of M41's brightest stars form a trapezoidal figure that bears a strong resemblance to the "keystone" asterism in Hercules. M41's other stars appear to form lines and curving arcs that extend away from the central trapezoid.

From edge to edge, M41 spans 38' — larger than the disk of the Full Moon. Because of this, low-power, wide-field eyepieces are best for taking in all of the cluster. High magnifications fail to provide a satisfying view of the entire group.

The 2nd-magnitude star Wesen (Delta Canis Majoris) marks a convenient starting point for finding a pair of open clusters. Just 1.5° northeast of Wesen, well within the same finderscope field, is **NGC 2354**. This cluster is composed of one hundred 9th-magnitude and fainter stars contained within 20'. Through binoculars, NGC 2354 appears as a faint mist nestled in a rich surrounding field. A 4-inch telescope resolves the cluster's brightest stars. With an instrument twice that size,

NGC 2243
Mag. 9.4   Size 5'   by Jim Barclay

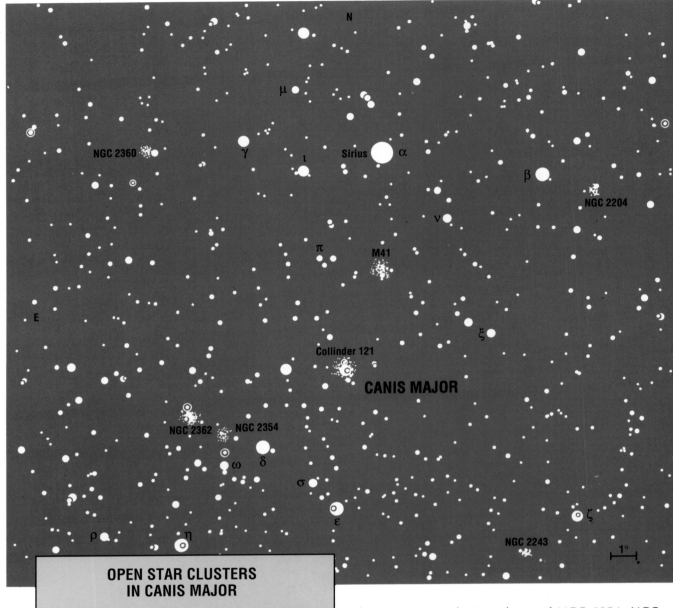

OPEN STAR CLUSTERS IN CANIS MAJOR

Object	R.A. (2000.0)	Dec.	Mag.	Size
NGC 2204	6h15.7m	-18°39'	8.6	13'
NGC 2243	6h29.8m	-31°17'	9.4	5'
M41 (NGC 2287)	6h47.0m	-20°44'	4.5	38'
Cr 121	6h54.2m	-24°38'	2.6	50'
NGC 2354	7h14.3m	-25°44'	6.5	20'
NGC 2360	7h17.8m	-15°37'	7.2	13'
NGC 2362	7h18.8m	-24°57'	4.1	8'

Object: Designation in Messier, NGC, or Collinder catalog. **R.A. and Dec.:** Right ascension and declination in equinox 2000.0. **Mag.:** V magnitude of cluster. **Size:** Diameter in arcminutes.

you'll see more than twice as many stars, with many of the brighter cluster members taking on a soft, yellowish hue. Once again, your widest-field eyepiece will yield the most pleasing views.

That rule does not hold true for our next object, however. Located 1° northeast of NGC 2354, **NGC 2362** consists of 60 stars crammed into 8'. Though small, NGC 2362 is a snap to spot since it surrounds the 4th-magnitude star Tau Canis Majoris. Large binoculars and small telescopes reveal a few 7th-magnitude cluster stars, along with Tau, embedded in the haze of fainter, unresolved stars. Eight- to 10-inch scopes display nearly all of the cluster's stars.

To find the next cluster, begin at 2nd-magnitude Murzam (Beta Canis Majoris), some 5° west-southwest of Sirius. From Murzam, move another 1.5° to the west-southwest, where your finder will uncover a single 6th-magnitude star and a faint glow nearby. The faint glow is the collective effect of the 80 stars belonging to **NGC 2204**. This cluster is less impressive than M41 or NGC 2362, but it is pretty enough to make the hunt worthwhile. You'll do best by switching to a moderate-power eyepiece first, because the cluster stars — none of which are brighter than 12th magnitude — all but hide in the rich background stars at low magnifications. The main body of the cluster measures 6' by 4' and outlying members extend this to 13'.

NGC 2354
Mag. 6.5 Size 20' by Jim Barclay

Easier to spot is **NGC 2360**, located about 8° east of Sirius and just east of a 7th-magnitude field star. Here you'll find a compact group of 80 stars that to many observers appears like an arrowhead. This shape is even discernible in large binoculars, although most of the stars are too faint to be seen individually. A good 4- to 6-inch telescope, however, cleanly resolves NGC 2360, showing many yellow and blue-white stars scattered across the cluster's 13' span.

Without a good view to the south, most observers tend to shy away from our next open cluster. **NGC 2243** lies in the constellation's southwestern corner and culminates less than 20° above the horizon for much of the United States, Canada, and Europe. Add to this its isolation from any bright stars, and it's easy to see why NGC 2243 is known to so few amateurs.

Actually, NGC 2243 is easier to find than it might appear at first. Begin your hunt at Adhara (Epsilon Canis Majoris) and cruise 7.5° southwest of 4th-magnitude Lambda Canis Majoris. NGC 2243 lies adjacent to a 6th-magnitude star, less than 2° to Lambda's north-northeast. Once in the region, your finderscope should reveal its small, subtle glow.

NGC 2243 is an exceedingly rich open cluster, in the same league as M11 in Scutum. The limited resolving power of small scopes causes it to look more like a globular cluster. An 8-inch instrument reveals only ten individual stars set within the group's hazy glow, while my 13.1-inch f/4.5 Newtonian doubles the star count. Because of its compactness, NGC 2243 can handle magnification well, so don't be afraid to crank it up!

Our last cluster belongs to the Collinder catalog of open clusters. This family of clusters is best known for large size and bright stars. **Collinder 121** spans 50' and shines at magnitude 2.6. Omicron[1] Canis Majoris, lying on the cluster's northern edge, shines at 4th magnitude and contributes much to the group's overall brightness. The other 20 or so stars in the cluster are not as bright — two shine at 6th magnitude and the rest between magnitudes 8 and 10. Combined with the starriness of the surroundings, the stars' magnitudes make it difficult to tell exactly where Collinder 121 begins and ends.

We've merely sampled the many fine open clusters scattered across Canis Major. With such a wide variety of sights from which to choose, don't let the cold get to you — conquer the beauty of Canis Major!

These objects are best visible in the winter evening sky

The Distant Suns

Journey to the outer limits of the Galaxy to observe gigantic spheres of ancient stars.

by David Higgins

Globular star clusters are some of the most beautiful and challenging deep-sky objects visible in amateur telescopes. Enormous spheres of ancient stars, they are extraordinarily large and bright when compared with, say, our solar system. Most globulars measure hundreds of light-years across and glow with the light of tens of thousands of Suns. Globular clusters are found in an enormous spherical halo surrounding the central bulge of the Galaxy. Over extremely long periods — tens of millions of years — the clusters orbit the center of the Galaxy in paths highly inclined to the galactic plane.

The unusual nature of these objects as semi-independent communities of stars outside the Galaxy's disk remained unknown until less than a century ago. Rather than condensing into the Galaxy's disk — the fate of most of the gas and dust that went into the Milky Way — globulars formed from blobs of material that escaped being pulled into the disk.

Today, twelve billion years later, they appear in the eyepiece as fuzzy disks of gray-green light surrounded by a peppering of faint, resolved stars. A given cluster's brightness, size, and the amount of detail you'll be able to resolve all depend on your telescope and the individual cluster you're observing. Globular clusters vary considerably in size and brightness. Additionally, these intrinsic differences appear exaggerated because the clusters lie at a wide range of distances. The clusters we'll explore lie at distances of 7,500 light-years to more than 110,000 light-years. Yet the views of globulars depend on more than just their distances: some are heavily obscured by intervening dust and gas.

But one thing is constant with all globulars: they let you reach back into the early days of the Galaxy and see some of the oldest objects known.

The first time you observe challenging globular clusters you might be struck by an apparent contradiction. If they're large, bright objects, why do they appear rather faint and challenging in the eyepiece? **NGC 5897** is a superb example of a challenging globular. Located in the dim constellation Libra the Scales, this 8.6-magnitude globular is large but has a low surface brightness. Measuring 12.6' across and lying at a distance of nearly 40,000 light-years, NGC 5897 is simple to find but not so simple to resolve into stars. To locate NGC 5897, start by aiming your telescope toward the 2nd-magnitude star 20 Librae. Next, draw an imaginary line northeast to 35 Librae. NGC 5897 lies about halfway between these two stars.

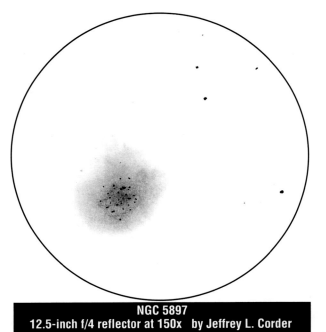

NGC 5897
12.5-inch f/4 reflector at 150x by Jeffrey L. Corder

In an 8-inch scope fitted with a wide-field eyepiece, NGC 5897 appears as a soft glow elongated east-west. You can resolve the edges of this cluster with moderate magnifications, but because NGC 5897 has a low surface brightness the entire cluster appears to vanish into the sky background if you use too much power. Larger scopes show NGC 5897 as a faint glow with some central brightening. The cluster's core is partially resolved at medium magnifications. Through a 17.5-inch or larger instrument NGC 5897 appears like a faint spiral galaxy. As you increase the magnification in a large scope, NGC 5897 appears fainter, but a handful of stars scattered across the core comes into view. Several strings of stars that heighten the spiral effect are visible extending east-west across the cluster.

Much closer to home than NGC 5897 lies **NGC 6539**, a cluster lying on the border between Serpens

NGC 5897
Mag. 8.6   Size 12.6'   Palomar Mountain Observatory photo

Challenging Globular Clusters

Object	R.A.(2000.0)	Dec.	Mag.	Size
NGC 5694	14h39.6m	-26°32'	10.2	3.6'
NGC 5897	15h17.4m	-21°01'	8.6	12.6'
NGC 6229	16h47.0m	+47°32'	9.4	4.5'
NGC 6517	18h01.8m	-8°58'	10.3:	4.3'
NGC 6539	18h04.8m	-7°35'	9.6:	6.9'
IC 1276	18h10.7m	-7°12'	?	7.1'
NGC 7006	21h01.5m	+16°11'	10.6	2.8'

Object: Designation in NGC or IC catalog. **R.A. and Dec.:** Coordinates in equinox 2000.0. **Mag.:** V magnitude of cluster. **Size:** Diameter in arcminutes.

Some globular clusters appear dim because we see them through the Milky Way's thick veil of dust

NGC 7006
Mag. 10.6   Size 2.8'   by Preston Scott Justis

Cauda and Ophiuchus. Although it's close — only 7,500 light-years away — this cluster is obscured by dust and appears fainter than NGC 5897. NGC 6539 suffers from low surface brightness. In small scopes it appears as a faint haze with no central brightening. In larger scopes, the unresolved haze appears irregular in shape. In a 17.5-incher the haze appears larger but is not resolved into stars. With averted vision the cluster takes on a grainy appearance — it's on the verge of resolution — across the core. A string of faint field stars is visible on the western edge. Low magnifications seem to give the best views of NGC 6539. High magnifications make the object appear too faint.

More than three times as distant as NGC 6539 is the faint and challenging globular cluster **NGC 6517**. Intrinsically brighter than either NGC 6539 or NGC 5897, NGC 6517 is relatively easy to find, despite its distance of 25,000 light-years. To locate NGC 6517, start at the 2nd-magnitude star 64 Ophiuchi. Extend an imaginary line to the northeast about 2° and you'll come to 69 Ophiuchi. NGC 6517 lies at the midpoint of this line. NGC 6517 measures 4.3' in diameter and glows at magnitude 10.3. In small scopes it gives the appearance of a small glowing ball with little central brightening.

Larger scopes do not improve the view much. The core appears brighter and a haze becomes visible surrounding the nucleus. In 17.5-inch or larger scopes, the haze surrounding the core becomes granular in appearance. Try high magnifications with this cluster.

Farther out lies the challenging cluster **IC 1276**, a very difficult object for small telescopes. Heavily obscured and faint, this cluster is far more challenging than NGC 6539 — which lies only 1.5° to the west. Glowing at magnitude 10.3 and measuring 7.1' in diameter, you would think that IC 1276 would not be that difficult of an object to find. Nothing could be farther from the truth. IC 1276 not only lies at a great distance from the Sun, 42,000 light-years, but it is also heavily obscured from all of the gas and dust clouds that lie between us and the cluster. In large telescopes, IC 1276 appears as a faint haze with no central brightening. Averted vision is required for a decent view on nights of other than good seeing. Use low magnifications and wide-field eyepieces to catch a glimpse of IC 1276. Small scope owners will achieve success by simply finding this elusive cluster.

The final three clusters are all very distant objects, all lying far out in the Galaxy's halo more than 100,000 light-years away. They are also approximately the same intrinsic brightness. However, they offer different degrees of difficulty because of varying amounts of interstellar gas and dust that block their light.

NGC 6229 is a small globular cluster in Hercules with a high surface brightness — despite its great distance of 102,000 light-years. Glowing at magnitude 9.4 and measuring 4.5' across, NGC 6229 is not difficult to locate. Start by finding 42 Herculis. Extend an imaginary line about 4.5° to the southeast and you'll come to the bright star 52 Herculis. The cluster lies about halfway along this line and a few arcminutes northeast.

In 6-inch or smaller scopes, NGC 6229 appears as a small, faint, fuzzy ball. Don't hesitate to use high magnification on this cluster, but don't expect to resolve it easily. Twelve-inch and larger scopes begin to resolve the brightest stars in this group. In 17.5-inch

NGC 6229
Mag. 9.4   Size 4.5'   by Jack Metcalfe

NGC 5897
Mag. 8.6   Size 12.6'   by Martin C. Germano

and larger scopes, NGC 6229 appears much brighter and the core seems like a gritty ball of sand surrounded by a faint haze. With averted vision you'll see some individual stars within this haze.

By using 20 Librae as a starting point we can track down Hydra's most challenging globular cluster, **NGC 5694**. This object lies 105,000 light-years away but is easier to see than NGC 6229 because of less interstellar absorption. From 20 Librae, move about 5° west and you'll find a string of 4th- and 5th-magnitude stars that extends roughly northwest-southeast and ends at the 3rd-magnitude star 58 Hydrae. Two degrees west of this string of stars lies NGC 5694. This small globular glows at magnitude 10.2 and is very concentrated so its surface brightness is fairly bright. NGC 5694 is bright enough that if you don't find it right away you can do some careful sweeping to spot it.

In telescopes 8-inches or smaller NGC 5694 appears as a small bright fuzzy ball. You can use high magnifications on this small cluster but don't expect to resolve it. In larger scopes the core is surrounded by a faint haze that appears somewhat granular in appearance. In the really large backyard scopes the core becomes relatively bright and the surrounding haze is much more granular. Averted vision will show several stars across the face of the cluster.

Delphinus the Dolphin contains **NGC 7006**, our final and most remote globular cluster. Glowing at magnitude 10.6 and measuring only 2.8' in diameter, NGC 7006 is a difficult object for small telescope users. Its apparent faintness is due to both a relatively large distance of 113,000 light-years and some obscuration by dust. Unfortunately no bright stars lie nearby to act as convenient signposts, so star hopping to NGC 7006 takes patience and determination.

The simplest way to find NGC 7006 is to center the 3rd-magnitude star 12 Delphini in a wide-field eyepiece. Move 3.5° east and you'll see NGC 7006 move into the eyepiece field. In 6-inch or smaller scopes NGC 7006 is a faint glow surrounding a somewhat brighter core. For some observers averted vision may be required to see this faint cluster. Scopes in the 12-inch range don't improve the view much although the core becomes a bit brighter and much more concentrated, and the haze that surrounds the core becomes easier to see. Backyard scopes 17.5 inches and larger show a much brighter haze surrounding a concentrated core. Averted vision makes the haze granular and at high powers you will see stars around the edge of the cluster but not across its face.

Each of these globular clusters presents a real challenge for small telescope observers. If you don't see them immediately, don't give up — simply try again on a darker night. As you successfully hunt down these globulars you'll gain experience with seeing faint objects and sharpen your observing skills. You'll also gain an overview of the structure of the Galaxy, a sort of mental, three-dimensional picture of how this vast halo of globulars surrounds the galactic disk. What better way to spend a warm summer evening than basking in the glow of our home Galaxy? □

These objects are best visible in the summer evening sky.

SPLIT A STAR IN TWO

Here are five close double stars that require extra observing skill to detect their binary nature.
by Alan Goldstein

Double stars have a special place in the realm of deep-sky observing. Unlike deep-sky objects, double stars are immune to light pollution and moonlight. They are available whenever you want to observe as long as the sky is clear. Best of all, slightly hazy evenings — useless for observing nebulae and galaxies — lend themselves perfectly to double star watching. On hazy evenings the atmospheric seeing is generally steady, which allows the starlight to pass through the atmosphere without distortion. The end result is clear, sharp views of close, colorful double stars.

Close double stars, those with components separated by less than five arcseconds, offer a real challenge to backyard observers. Some of these pairs have great magnitude differences between their stars, the secondaries glowing feebly in comparison to the primaries. This difference makes the faint secondary extremely difficult to spot because of glare from the primary star.

The brightest star in the sky, **Sirius** (Alpha [α] Canis Majoris), is a close double that offers an enormous contrast between its components' brightnesses. Sirius A shines brilliantly at magnitude −1.5, a white beacon that seems like a searchlight in any telescope. Sirius B, the companion, lies about 4″ due north of Sirius A and is a magnitude 8.4 star — about 4,000 times fainter than Sirius A. To see the faint secondary, place an occulting bar — a small strip of cardboard — inside the eyepiece at the focal plane. This will block enough light so that if seeing conditions are steady you should spot Sirius B.

Sirius A is several times more massive and 23 times more luminous than our Sun. Its stark white color results from a surface temperature of 12,000 K and a spectral type of A1. In contrast, Sirius B is a white dwarf, a star with a mass comparable to the Sun's that is compressed into a white-hot sphere only 19,000 miles across.

Reasonably close to Sirius in the sky is **Rigel** (Beta [β] Orionis). The seventh brightest star in the sky, Rigel shines at magnitude 0.1. Rigel is a double star with a secondary that glows at magnitude 6.7, considerably brighter than Sirius B. Observing this star is not as difficult as finding Sirius B, although it is still a challenge. Look for Rigel B with a high-power eyepiece some 9″ south of Rigel A. As with Sirius, you may have to use an occulting bar to block light from the primary.

Both components of Rigel are visible without difficulty in a 4-inch refractor. In large telescopes Rigel is one of the finest double stars in the winter sky; both stars show a slight bluish tint. Rigel B itself is a close binary, never exceeding a separation of 0.2″. When observing with large telescopes, some observers have reported Rigel B to be elongated. Can you see anything unusual when looking at Rigel B?

The similarity between Rigel A and B and Sirius A and B is purely visual. Rigel is one of the most luminous stars known in our Galaxy, over 55,000 times brighter than the Sun. In our sky Rigel is 1.5 magnitudes fainter than Sirius, yet Rigel lies 900 light-years away, more than 100 times more distant than Sirius. Rigel B lies more than 200 billion miles from Rigel. At that distance you would have to observe Rigel A and Rigel B for over two hundred years to detect any orbital motion.

Castor (Alpha Geminorum) is another challenging star for double star observers. Castor A and B shine at magnitudes 2.0 and 2.8. The stars are separated by a mere 2″, which makes splitting this pair rather challenging. The system reached periastron in 1969 (1.8″) and will slowly widen to 6.5″ over the next 110 years. The orbital path is inclined 25° from the edge-on position, making the two components much more difficult to observe when their orbital motion carries one star

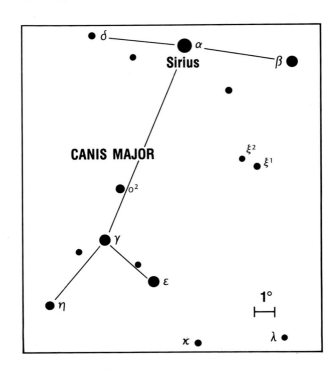

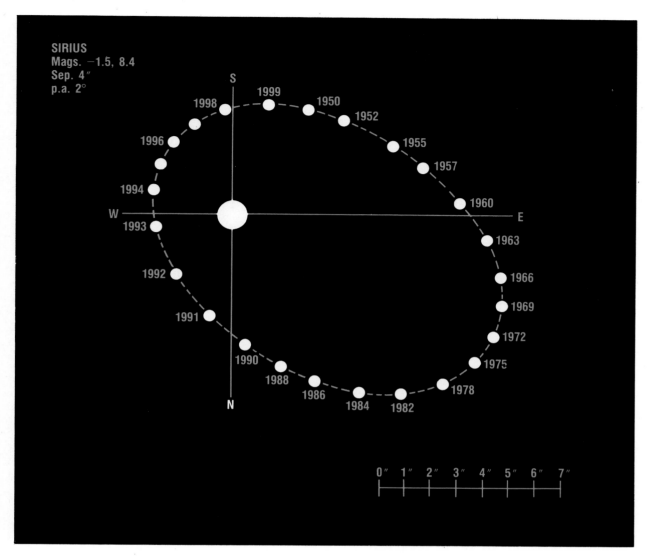

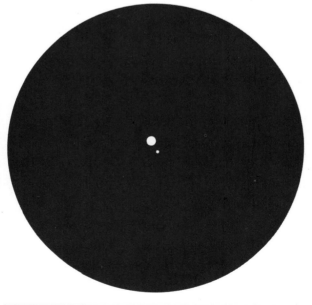

Sirius
8-inch f/10 SCT at 200x
by David J. Eicher

THE BRIGHTEST STAR IN THE SKY, Sirius is a close double with an 8th-magnitude secondary. Splitting Sirius requires a large telescope and an occulting bar placed in the eyepiece.

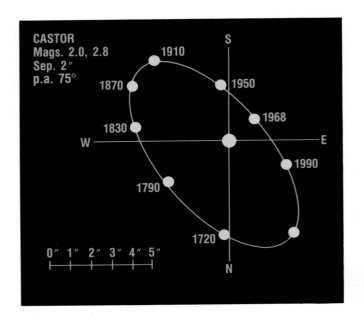

Castor
3-inch f/10 Reflector at 120x
by Glenn F. Chaple, Jr.

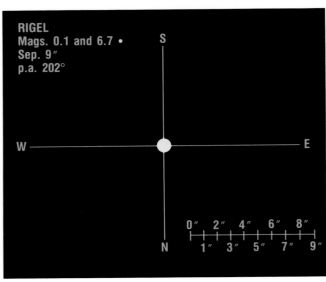

Rigel
3-inch f/10 Reflector at 120x
by Glenn F. Chaple, Jr.

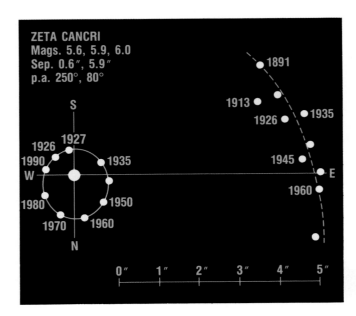

Zeta Cancri
6-inch f/8 Reflector at 200x
by James Mullaney

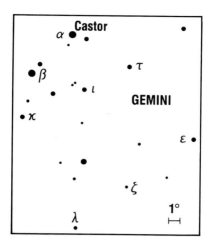

Close, Challenging Doubles

Object	R.A. (2000.0) Dec.		Mag.	Sep.	P.A.
Rigel	5h14.5m	−8°12′	0.1, 6.7	9″	202°
Sirius	6h45.1m	−16°43′	−1.5, 8.4	4″	2°
Castor	7h34.6m	+31°53	2.0, 2.8	2″	75°
Zeta Cancri	8h12.2m	+17°39′	5.6, 5.9, 6.0	0.6″, 5.9″	250°, 80°

Object: Star name. **R.A. and Dec.:** Right ascension and declination in equinox 2000.0. **Mag.:** V magnitudes of component stars. **Sep.:** Separation of components in arcseconds. **P.A.:** Position angle in degrees.

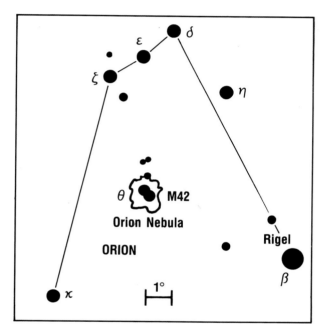

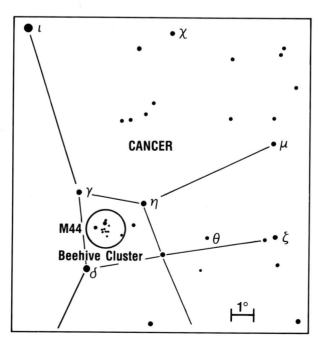

nearly in front of the other. Lying 72.5″ distant in position angle 164° is a third member of the Castor system. At magnitude 9.1, this red dwarf lies more than 100 billion miles from the other stars and has an orbital period exceeding 10,000 years.

A triple star system is unusual, but Castor does not stop there. Each member of the Castor system is also a binary, too close to be resolved with optical telescopes. Castor A² has an eccentric orbit around A¹ and completes one revolution every nine days. Castor B's components revolve in a three-day circular orbit. Castor C, also known as YY Geminorum, is an eclipsing binary. The stars eclipse each other every 9.8 hours, causing a measurable drop in the system's brightness.

Our last challenging star is **Zeta Cancri**, located about 5° west and slightly south of the Beehive Cluster. To the unaided eye, Zeta Cancri is simply an inconspicuous 5th-magnitude star. However, a 10-inch scope reveals that Zeta Cancri is a triple system. All three stars have a yellowish cast due to similar spectral types of F8, G0, and G2. The A and B components shine at magnitudes 5.6 and 5.9 with a separation of just 0.6″. At the computed distance of 71 light-years, the stars have a real separation of 1.8 billion miles — about the distance between the Sun and Uranus.

Zeta Cancri's third component (Zeta C) is a 6th-magnitude star lying 5.9″ distant in position angle 80°. Its quick orbital motion around the A-B pair has carried it through a position angle change of more than 70° over the past 150 years. Zeta C is also no loner: it is an astrometric binary, a star detectable as a double only by precise measurements of its motion in the sky. Zeta C's companion lies at a distance of about 500 million miles and is too faint to see visually, suggesting that it may be a white dwarf star or another object with an extremely low luminosity.

These four challenges merely represent the tip of the iceberg in double star observing. Thousands of challenging doubles — close pairs with striking differences in brightnesses and colors — are visible in backyard telescopes. Spend some time with these objects and you'll find that splitting the stars is a lot of fun. □

These objects are best viewed at moderately high power on nights of steady seeing.

Five Challenging Globulars

Globular star clusters test your observing skills by making you perceive slight differences in star density, resolution, and brightness.
by David Higgins

Objects predating the formation of our galaxy, globular star clusters are the ancient denizens of the universe. Composed of very old stars, globulars float like bees around the great central hub of the Milky Way in an irregular halo. In marked contrast to open star clusters, which contain up to several thousand stars, globulars pack in hundreds of thousands or millions of stars.

Because of the numerous red giant stars in globulars, most of these clusters are high surface brightness objects and are therefore easily visible in small telescopes. Some globulars, however, pose challenges for small scope observers. The reason may be low star density, a tightly compressed (and unresolvable) center, or simply a small apparent size. A handful of springtime globulars stand out as lures for those seeking a challenge.

The brightest and largest of the springtime globulars is **M3** (NGC 5272) in Canes Venatici. Shining at magnitude 6.4 and measuring 16.2' across, M3 ranks as the third brightest globular cluster in the northern sky. Estimated to be 10 billion years old, M3 is more than twice as old as our solar system. Generations of star formation and periodic dips through the plane of the Galaxy have removed virtually all of the gas and dust from this cluster, leaving a shell of old, yellow stars with empty space between them.

M3 is located in the southernmost portion of Canes Venatici. The easiest way to find it is to start by locating the 5th-magnitude star 9 Bootis. From there, sweep 2.5° northwest toward Beta Comae Berenices. On a dark night the fuzzy glow of M3 will be visible in any finder scope. If you're using a telescope smaller than 6-inches in aperture M3's core will not appear resolved. Instead, the cluster will resemble a large ball of gray-green light surrounding a bright, whitish core.

In a 6-inch telescope some resolution is possible, but the cluster's distinctive chains of stars are difficult to see with a 6-inch scope. A 12-inch telescope shows several strings of stars extending to the west of the cluster. Additional strings can be seen extending to the north and south. With high magnification you can resolve M3 almost to the core. The western half of the cluster is brighter than the eastern half.

Large telescopes show M3 quite differently. With high magnification the cluster's core is resolved into hundreds of stars spread uniformly across the face of the cluster. The strings of stars that extend to the north

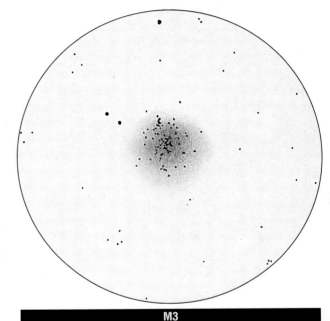

M3
8-inch f/10 SCT at 50x   Sketch by David J. Eicher

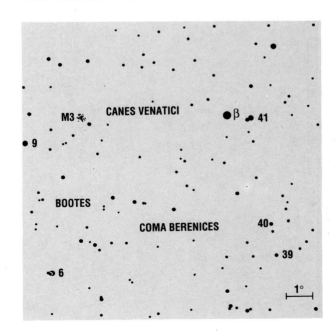

M3
Mag. 6.4 Size 16.2' by Preston S. Justis

M53
Mag. 7.7   Size 12.6'   by Jack Newton

NGC 5053
Mag. 9.8   Size 10.5'   by Martin C. Germano

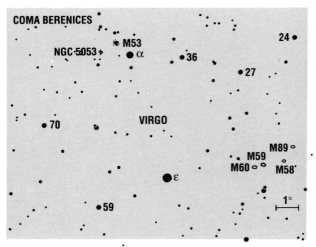

and south are much more pronounced. The strings extending to the west are brighter and more noticeable. One in particular extends about 5' to the west and then suddenly makes a 90° bend to the north, coming to an end at a bright field star. Telescopes in the 17-inch class and larger give the cluster the appearance of being extended north and south. Several loops of stars can be seen extending north and south of the cluster, but the eastern edge lacks star chains or loops.

Large and loose **NGC 5466** in Bootes is very faint and has an extremely low surface brightness. Although it glows at 9th magnitude and measures 11' across, the stars in NGC 5466 are scattered and sparse. Because the group lacks a bright central condensation, you have to see some of its brightest individual stars in order to view the cluster. To hunt it down start back at 9 Bootis but this time sweep 2° to the northeast. Use a low-power, wide-field eyepiece.

Small telescopes show NGC 5466 only as a faint haze. Little structure or detail is visible even at high magnifications. With a 6-inch scope at 100x NGC 5466 appears like a weakly glowing disk that stands out only marginally from the sky background. Higher powers reveal a hint of resolution. At 200x the amorphous glow seems peppered with knots that during moments of extraordinarily good seeing snap into focus as little groups of pinpoint stars. The effect is visible only with averted vision, however. A 10-inch telescope is required to routinely resolve the cluster's brightest members. With this aperture NGC 5466 appears as a disk of greenish light with a dozen or more tiny stars suspended in the glow. In a 12-inch scope the view is more distinct. At low magnifications the cluster's glow stands out well from the sky background. At high magnifications two dozen or more stars are visible, including a prominent string of stars

NGC 5466
Mag. 9.1   Size 11.0'   by Martin C. Germano

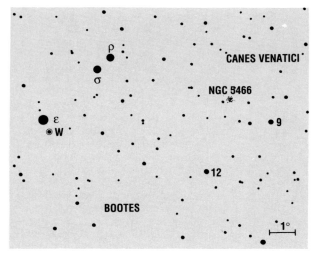

Challenging Spring Globulars

Object	R.A.(2000.0)Dec.	Mag.	Size
NGC 4147	12h10.1m +18°33'	10.3	4.0'
M53 (NGC 5024)	13h12.9m +18°10'	7.7	12.6'
NGC 5053	13h16.4m +17°42'	9.8	10.5'
M3 (NGC 5272)	13h42.2m +28°23'	6.4	16.2'
NGC 5466	14h05.5m +28°32'	9.1	11.0'

Object: Designation in Messier or NGC catalog. **R.A. and Dec.:** Right ascension and declination in equinox 2000.0. **Mag.:** V magnitude of cluster. **Size:** Diameter in arcminutes.

that extends outward from the cluster's eastern edge.

Coma Berenices contains three spring globular clusters. **M53** (NGC 5024) is the largest and brightest of the three. Although not as large or impressive as M3, it is nearly as bright. Easily visible in a finder scope, M53 can be found by starting at Alpha Comae, the brightest star in Coma Berenices. Using low magnification, sweep 1° to the northeast and M53 will show up in the field of view. In small telescopes the cluster appears as a large, bright cometlike ball with a bright middle that gradually fades outward. In a 6-inch scope it appears partially resolved around the edges and shows a sprinkling of stars. A 12-inch scope at high magnification resolves part of the core into a mass of individual stars. At lower magnifications a chain of stars on the eastern edge extends north and south.

In a 17.5-inch scope at high power M53 becomes a mass of twinkling fireflies. The core is resolved with strings of stars extending outward from the northern and western edges. A few loops of stars are visible at low magnifications. A solitary bright star floats on the northern edge of the cluster.

A faint neighbor of M53 offers an extreme challenge. **NGC 5053** lies about 1° southeast of M53 and can be glimpsed as a faint hazy spot in telescopes equipped with low-power, wide-field eyepieces. Don't expect to see this one in a finder scope. NGC 5053 glows at magnitude 9.8 and measures 10.5' across. In 6-inch or smaller telescopes the cluster is hard to find and does not offer any detail other than a faint haze. A 12.5-inch scope provides a view of NGC 5053 that almost resembles an extremely rich open cluster, with a sprinkling of stars across the face of the cluster. Large telescopes show a central concentration of stars. The cluster is irregular in shape and gives the impression of a misshapen triangle oriented roughly east-west.

Our final challenging globular lies in the corner of Coma Berenices. **NGC 4147** is situated on the northern edge of the Virgo Cluster of galaxies, causing some confusion about its identity when viewed with a finder scope. To find this faint cluster start with Beta Leonis and 93 Leonis. From a point midway between the two stars extend a line eastward for about 8°. Now you will see the 5th-magnitude star 11 Comae Berenices. These three stars form a triangle that is easy to see from a dark-sky site. From 11 Comae Berenices extend a line northwestward 2.5° degrees toward 93 Leonis. This point marks the location of NGC 4147.

NGC 4147 is very small compared to M3. NGC 4147 shines at magnitude 10.3 and measures 4' across. Small telescopes do not resolve the core but they do show a bright central region that gradually fades outward. A 12-inch scope doesn't resolve the core either but some stars are visible on the cluster's edges. In larger scopes the core appears starlike due to its great brightness. The cluster's edges are resolved into stars although no strings or loops are visible.

Who says springtime observing has to mean galaxies? Try a change of pace by observing some objects that belong to the good ol' Milky Way. Although the spring sky holds fewer globulars than the richer region of the summer Milky Way, the springtime clusters challenge your observing skills. Try them on the next clear night and you'll be amazed at the variety and beauty of these ancient celestial gems. ☐

These objects are best viewed at moderately high power on nights of steady seeing.

Gazer's Gazette

The Secret World of Dark Nebulae

by David Higgins and David J. Eicher

If you walk outside and look up on a clear autumn night in suburbia, you'll likely see a few dozen bright stars. After dark-adapting your eyes for several minutes, you'll see hundreds of stars and, perhaps, a faint glow marking the Milky Way.

Walk outside under a black, rural sky, though, and the view is dramatically different. Against a sky full of stars you'll see a bright, detailed Milky Way that almost looks like an airbrush painting. Under these conditions you can see not only bright patches but dark clumps and filaments. These dark areas are dark nebulae, clouds of opaque dust grains in space that block starlight from behind.

Dark nebulae are not difficult to observe, provided your observing location and equipment meet a few basic requirements. Because dark nebulae are defined by bright objects surrounding them, the sky must be reasonably dark. Your telescope is an important consideration. The more aperture you have, the better: mirrors in the 12-inch or larger class reveal faint stars and therefore make seeing hundreds of dark objects possible. You don't need a large scope, however. Several dozen fine examples of dark nebulae are easily visible in binoculars or small telescopes under a dark sky.

The large, dark rifts in the Milky Way have been observed since antiquity. Telescopic dark nebulae, the smaller patches and lanes of cosmic dust that in groups form naked-eye dark nebulae, have been observed since the invention of large telescopes about two hundred years ago. In 1784 the English observer Sir William Herschel studied what is now called the Rho (ϱ) Ophiuchi dust cloud near the bright star Antares. Herschel observed many lanes and patches where the overall star density was low relative to the surrounding starfields. Herschel couldn't determine the nature of dark nebulae, but he diligently cataloged many of these objects. A detailed study had to wait until photography revealed the full extent of dark nebulae.

In 1884 Edward Emerson Barnard, a

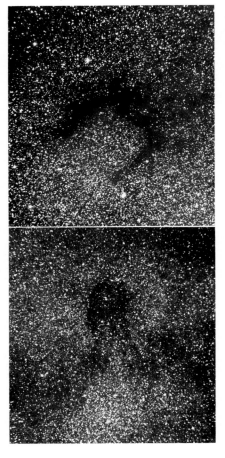

Top: Dark nebulae are easier to see than you may think. B143 is an opaque, C-shaped cloud in Aquila.

Above: B87 in Sagittarius is known as the "Parrot's Head Nebula." Photos by Martin C. Germano.

Opposite page, top: The Lagoon Nebula (M-8) in Sagittarius contains a large amount of dark nebulosity. Photo by Jack Newton.

Opposite page, bottom: A short exposure reveals many dark lanes around the hourglass-shaped nebula near M-8's center. Photo by Jack B. Marling.

professor of astronomy at the University of Chicago, began making long-exposure photographs of the Milky Way. For twenty-five years he photographed and studied the Galaxy and gradually came to believe its many dark lanes were pockets of obscuring material. The issue was finally settled in 1930 when Robert J. Trumpler studied the angular sizes of star clusters. Trumpler knew from previous research that the smaller the open cluster, the dimmer its stars appear. This holds true because most open clusters are relatively equal in size. However, Trumpler found that stars in many distant open clusters were much too faint to follow the formula. Thus, something was absorbing some of their light, something that lay between the cluster and us. The dimming Trumpler observed was caused by interstellar dust.

When Barnard cataloged dark nebulae, he gave them numbers, preceded by a "B" designation (for Barnard), similar to the "M" designation Charles Messier assigned nebulae and clusters in the eighteenth century. In 1927 Barnard published his catalog of 370 dark nebulae in a two-volume classic called *A Photographic Atlas of Selected Regions of the Milky Way*. The first volume contains photographic prints of regions photographed by Barnard, and the other contains descriptions of the objects and charts that correspond to the photos.

Because the photographs in this work are actual photographic prints pasted in by hand, the book was very expensive to produce, so less than a thousand copies were made. Today it is a rare item, but it can be found in university and city libraries and is sometimes offered for sale by used book dealers. Barnard's *Selected Regions* gives sizes, shapes, positions, and descriptions for his catalog of dark nebulae. The charts are easy to use because they show bright stars and NGC objects, which serve as guideposts. Although some of the nebulae visible on Barnard's photos are beyond the reach of amateur telescopes, many of these nebulae can be spotted by keen-eyed observers.

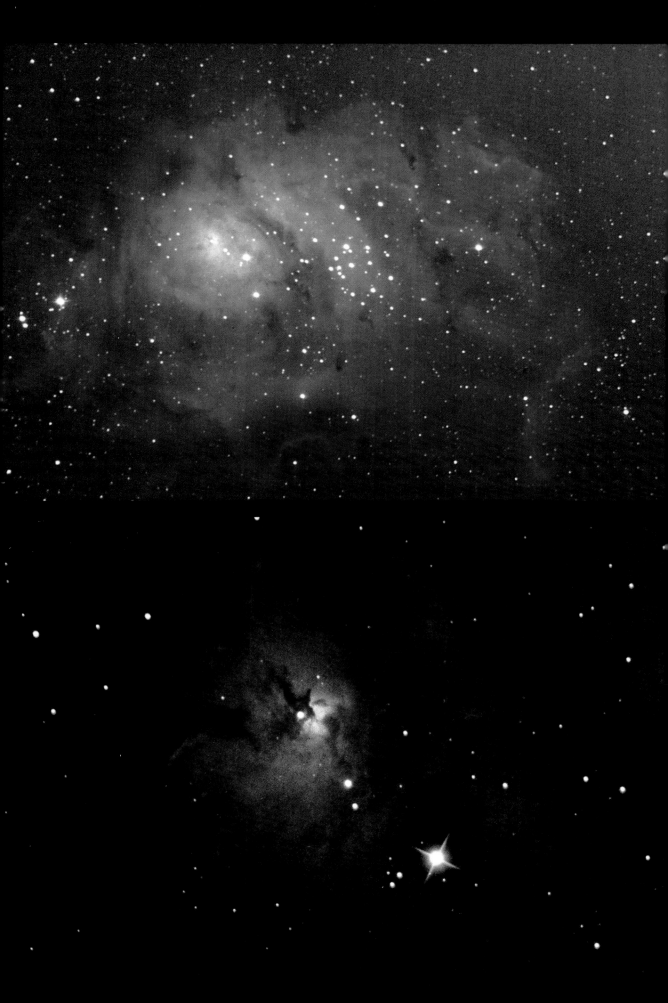

The best accessible list of dark nebulae for backyard observers appears in *Sky Catalogue 2000.0, Volume 2*. The dark nebula section lists 150 objects, giving epoch 2000 coordinate centers and brief descriptions. Opacity, or relative darkness, is also provided on a scale of 1 to 6, where 1 is the least opaque and 6 the most opaque. Two star atlases that show dark nebulae are Antonin Becvar's *Atlas of the Heavens* (now out of print but occasionally available from booksellers) and the just-released *Uranometria 2000.0*.

Dark Nebulae in the Autumn Sky

The following list of twenty-one of the best dark nebulae ranges from very easy objects to moderately difficult ones. Each dark nebula has its own distinctive characteristics. The list covers five bright early autumn constellations, all of which have bright stars that serve as guideposts for locating the nebulae.

What will a dark nebula look like in the eyepiece? It depends on the telescope, sky, and individual nebula. Some dark nebulae show well-defined boundaries; others merge gradually into the surrounding star clouds. Generally, dark nebulae appear as "blank areas" that show relatively few stars in an area otherwise packed with hundreds of stars. Some of the most dramatic examples are very opaque objects lying squarely in the plane of the Milky Way. These objects really do look like black patches or tunnels with no stars inside them. They are bordered by dense areas of stars, and the contrast is striking.

Aquila

Barnard 142 and *B143*. Both of these nebulae lie about 1° west of the bright star Gamma (γ) Aquilae and only 3° northwest of Altair. B142 forms an irregular horseshoe shape. B143 has a well-defined western edge and two narrow streamers 30' long extending to the west. The eastern edge shows no sharp outline. When observing these dark nebulae, you'll need a wide field of view.

B337. This very dark nebula lies about 2° northwest of B143 and is oval in shape. It measures 3' in diameter and contains very few stars. Although only a few stars are associated with B337, two fairly bright foreground stars lie in the middle of the nebula and help you identify the nebula.

Left: Sagittarius is a rewarding constellation for dark nebula hunters. Visible to the naked eye under a country sky, the "Great Galactic Dark Horse" is one of the Milky Way's largest and most magnificent dark structures. The Dark Horse extends through parts of Ophiuchus, Sagittarius, and Scorpius, covering thick Milky Way fields toward the galactic center. Photo by Al Lilge.

Inset: Under the Horse's belly lies the 30'-diameter dark nebula B172, also known as the "S Nebula." With a low-power telescope the S-shape is rather easy to see on a dark, moonless night. Photo by Ronald Royer.

Above: Sagittarius contains two of the sky's best and easiest to find dark nebulae. Located on the northeastern side of the bright star cloud M-24, B92 (center) can be seen with a 3-inch telescope. A solitary 12th-magnitude foreground star is the only object "within" B92. B93 (right center) is a larger, more oval nebulosity that is also easy to see with a small telescope. Photo by Scott Hicks.

Left: Four degrees south of the Lagoon Nebula lies the open cluster NGC-6520, a bright group of sixty stars bordered by dark nebulae to the north and east. The eastern nebula is B86, one of the most opaque dark nebulae in the Milky Way. This dark cloud's density and location in a rich starfield make it easy to see. Photo by Jack B. Marling.

B337 also has a narrow extension to the northwest.

Cepheus

B169, B170, and *B171.* All of these nebulae lie a degree northwest of Zeta Cephei. B169 is shaped like an elliptical ring about a degree wide and encloses a small group of stars. The northern portion is distinct, though the southern portion is rather ill-defined. B170 lies on the northwest edge of B169 and is an irregularly shaped strip spanning 26' by 4'. It is very opaque, showing no stars whatsoever across its surface. B171 lies on the eastern edge of B169 and is an irregular area of nebulosity about 19' long. This nebula is easily visible in small telescopes and shows no stars inside its boundary.

Cygnus

B145. Located about 1° northeast of 25 Cygni, B145 is a triangle-shaped nebulosity covered by a number of faint stars. It is bordered by two bright foreground stars, one on its northern edge and another on its eastern edge. The eastern edge is more opaque than the rest of the nebula. B145 measures about 45' long and is oriented east-west.

B356. This large, oval dark nebula lies about halfway between 59 and 60 Cygni. The nebula's eastern edge is more opaque than other parts. There are some field stars scattered across its surface; a bright field star is visible on the southeastern edge.

B363. B363 is a complex of gas and dust measuring 40' across, located just 1° northwest of the scattered open cluster M-39. The western end of B363 is larger than the eastern end, but the eastern edge has two extensions easily visible in small scopes. A bright field star lies on the southwestern edge, but few stars are visible across the nebula.

Ophiuchus

B64. This nebula lies only 30' west of the bright globular cluster M-9. B64 measures 20' across and is oriented north-south. This extension has an opaque northern section and a wide, less well-defined southern section.

B70 and *B74.* Both of these nebulae lie in the same wide-field eyepiece. B74 lies about 30' west of 44 Ophiuchi, measures 15' by 10', and is elongated north-south. It is slightly curved and has an extension on its northern side; the western and southern sides are the most opaque. B70 lies about 30' northwest of B74 and is much smaller, measuring 4' across.

B72. This is one of Barnard's most famous objects, the spectacular "S Nebula." It measures 30' across, so you'll need a wide field of view to see the "S" shape. B72 lies 30' north of B70 and is only 2' to 3' wide.

Sagittarius

B85. Most backyard observers have looked at the Trifid Nebula (M-20), but few realize that the Trifid's famous dark lanes are cataloged as a dark nebula, Barnard 85. This nebula is visible in small telescopes under a dark sky and appears detailed in telescopes with 8 inches or more aperture.

B86. A fine dark nebula is located in the Sagittarius star cloud, just east of the open star cluster NGC-6520. B86 is almost square-shaped, measures about 5' across, and has a high opacity that makes it easy to spot. With low and medium magnifications you'll see both the cluster and the nebula in the same field of view, which provides a wonderful contrast between darkness and light.

B87. Nicknamed the Parrot's Head Nebula, this large and roughly oval object lies about midway between Gamma (γ) Sagittarii and the brilliant open cluster M-7 in Scorpius. B87 measures 12' in diameter and is surrounded by several other dark nebulae visible in small scopes. A few stars shine through the dark nebulosity.

B88 and *B89.* Both of these nebulae lie inside a well-observed Messier object, the Lagoon Nebula (M-8). With a large telescope B88 is visible along the southern portion of M-8 and has a wispy appearance. B89 is a narrow dust lane that runs almost completely across the face of M-8. It measures roughly 2' by 0.5' and is easily visible even in small telescopes. Next time you observe the Lagoon, try using a UHC filter: it brings out the dust lanes extraordinarily well.

B92. B92 is another dark nebula associated with a Messier object: it lies directly in front of the brightest part of star cloud M-24. The dark cloud measures 15' by 9' and contains a single, 12th-magnitude star in its center. B92 is one of the best-defined dark nebulae because it is surrounded by thousands of stars in the dense Sagittarius region of the Milky Way.

B93. Lying about 20' northeast of B92 is the dark nebula B93. This well-defined object measures 6' by 2' across but is connected on its southern end to a dark nebulosity 15' long. B93 is very opaque and easily seen, but the southern extension is irregularly shaped and not as easily discernible. The northern end of B93 is slightly rounded and more opaque than the southern end, which makes the northern end look like a dark comet.

B289. This large dark nebula lies 2° southwest of B93 and measures a whopping 35' by 7' across. It is oriented north-south and requires a wide field and north-south sweeping to spot.

Dark nebulae provide a wonderful escape from the tried-and-true showpiece objects like the Orion Nebula and the Andromeda Galaxy. The next time you are out under a dark sky, take on the challenge of seeing some patches of dark cosmic dust.

Summer Dark Nebulae

Object	R.A. (2000) Dec.		Con.	Opacity	Size
B244	17h10.1m	−28°24'	Oph	5	30' x 20'
B250	17h13.1m	−28°21'	Oph	5	15'
B64	17h17.2m	−18°32'	Oph	5	20'
B70	17h23.5m	−23°58'	Oph	4	4'
B72	17h23.5m	−23°38'	Oph	6	30' x 3'
B74	17h25.2m	−24°12'	Oph	5	15' x 10'
B289	17h56.6m	−29°01'	Sgr	5	7' x 35'
B85	18h02.6m	−23°02'	Sgr	6	—
B86	18h02.7m	−27°50'	Sgr	6	4'
B88	18h03.8m	−24°23'	Sgr	5	2.7' x 0.5'
B89	18h03.8m	−24°23'	Sgr	6	2' x 0.5'
B87	18h14.3m	−32°30'	Sgr	5	12'
B92	18h15.5m	−18°11'	Sgr	6	12' x 6'
B93	18h16.9m	−18°04'	Sgr	5	12' x 2'
B337	19h36.8m	+12°27'	Aql	4	3'
B340	19h49.1m	+01°25'	Aql	5	7'
B142	19h40.7m	+10°57'	Aql	6	40'
B143	19h40.7m	+10°57'	Aql	6	30'
B145	20h03.5m	+86°02'	Cyg	4	35' x 6'
B356	21h01.9m	+06°31'	Cyg	4	24'
B363	21h26.6m	+08°36'	Cyg	5	40' x 7'
B169	21h58.9m	+58°45'	Cep	5	60'
B170	21h58.9m	+58°45'	Cep	5	26' x 4'
B171	21h58.9m	+58°45'	Cep	5	19'

These objects are best visible in the summer evening sky.

The Art of Observing Planetaries

The strategy for observing planetary nebulae is to search with low magnification and then switch to higher magnifications to see details.
by David J. Eicher

Planetary nebulae are a pleasant contradiction. Their fuzzy disks are physically similar but appear diverse in the eyepiece, which gives observers a nice variety of shapes, sizes, and brightnesses. Because these objects are spread throughout the Galaxy, some planetaries are easy to see with small telescopes. Others, however, challenge the largest, most sophisticated backyard instruments.

The end of a star's life, a planetary nebula is a shell of gas gently expelled from a star after it exhausts its fuel supply. Planetary nebulae are transitory objects: they slowly expand over the course of about 10,000 years before thinning imperceptibly into the surrounding space. All that is left is a dim, dwarf central star that slowly burns out like a glowing ember. However, only stars of average mass — including our Sun — will become planetary nebulae. Massive stars end their lives explosively as supernovae; tiny stars simply cool off.

One of the most challenging planetary nebulae for small scopes is **M97** (NGC 3587) in Ursa Major. M97 measures slightly over 3' across and has a brightness of magnitude 12.0. Because its dim light is spread over a fairly large area, M97 has a notoriously low surface brightness. This planetary nebula is much harder to see than the statistics suggest. In fact, most observers place M97 among the most challenging of all the Messier objects. Because of two dark areas that look like eyes on the surface of its disk, M97 is popularly known as the Owl Nebula.

Fortunately M97's location is easy to find. Start by aiming your finder scope toward Merak (Beta [β] Ursae Majoris), the southwestern star in the bowl of the Big Dipper. Swing your finder slightly over 2° southeast and you will come upon a pair of bright stars aligned north-south. The Owl Nebula lies 20' northeast of the solitary 6th-magnitude southern star.

Once you've located M97, what you'll see depends on the telescope you're using. A 3-inch scope is capable of showing the Owl as a tiny, misty spot of gray light but only at high magnifications on exceptionally transparent nights.

With their greater light-gathering power and correspondingly smaller fields of view, 6-inch and 8-inch scopes show a much brighter Owl Nebula. In an 8-inch scope at 170x the Owl looks like a 3'-diameter gray-green patch of light that is perfectly round and almost imperceptibly dimmer toward its edges. The field is strewn with faint stars invisible to smaller scopes, six of which form a distinctive semicircle around the Owl Nebula. This little feature is visible as a line of stars originating south of the nebula and circling clockwise to the brightest star, which lies just north of the Owl.

Ten-inch or larger scopes coupled with steady seeing present two curious features of the Owl. Using averted vision — glancing off to the side of the field — you will see the 12th-magnitude central star embedded squarely in the nebula's center. This tiny dot of light is responsible for shedding the layers of gas that formed the nebula. In addition, a 10-inch scope and averted vision show the two dark patches that gave rise to the nebula's popular name.

Two observing techniques will dramatically improve the detail you can see in planetary nebulae. First, gently moving the telescope in right ascension or declination by small amounts as you use averted vision makes it easier to see minute details. This is because your eyes are greatly sensitive to motion. Second, try using a red-transmissive nebula filter like the UHC filter made by Lumicon. This device transmits the nebula's light to your eye but blocks scattered terrestrial lighting. Using a nebula filter is particularly effective for spotting details like the Owl's eyes.

Southern Hemisphere observers have a nicely detailed planetary nebula in **NGC 3132**. Located in the constellation Vela at −40° declination, this planetary is invisible north of +50° latitude. Those who can observe Vela will find NGC 3132 by aiming a finder scope toward the 4th-magnitude star q Velorum, moving 2°

Challenging Springtime Planetaries

Object	R.A. (2000.0) Dec.		Mag.	Size	Tirion	Uranometria
NGC 3132	10h07.7m	−40°26'	8.2p	>47"	20	II: 399
NGC 3242	10h24.8m	−18°38'	8.6p	16"/1250"	13, 20	II: 325
M97	11h14.8m	+55°01'	12.0p	194"	2	I: 46
NGC 4361	12h24.5m	−18°48'	10.3p	45"/110"	14, 21	II: 328
NGC 6210	16h44.5m	+23°49'	9.3p	>14"	8	I: 156

Object: Designation in Messier or NGC catalog. **R.A. and Dec.:** Right ascension and declination in equinox 2000.0. **Mag.:** Photographic magnitude. **Size:** Diameter in arcseconds. **Tirion:** Chart on which object appears in *Sky Atlas 2000.0* by Wil Tirion. **Uranometria:** Chart on which object appears in *Uranometria 2000.0* by Wil Tirion, Barry Rappaport, and George Lovi.

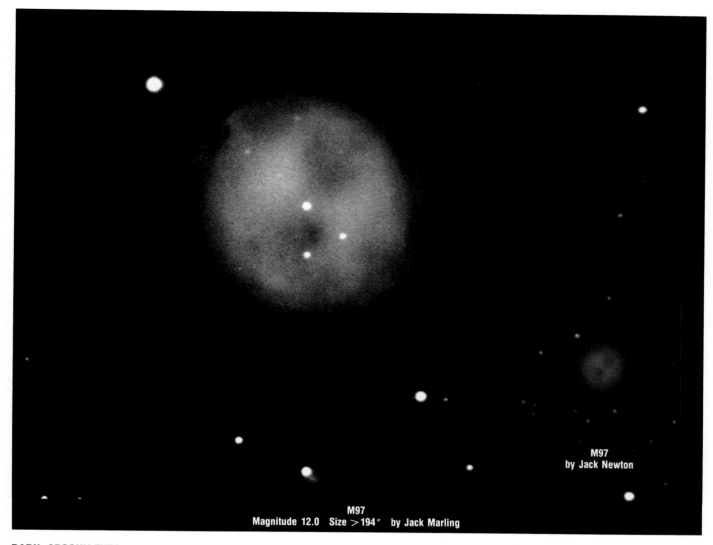

M97
Magnitude 12.0 Size >194″ by Jack Marling

M97
by Jack Newton

DARK, SPOOKY EYES gaze toward Earth from the Owl Nebula in Ursa Major. The remnants of gas expelled by a star that died suddenly, the Owl Nebula shows a planetary nebula's characteristic strong blue-green color.

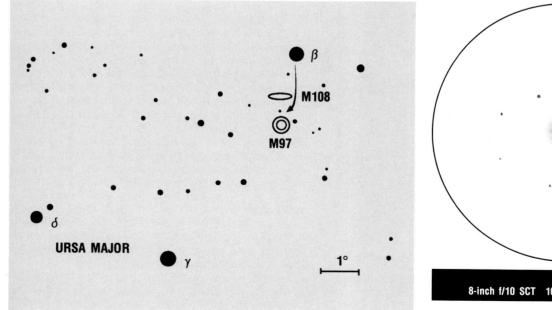

M97
8-inch f/10 SCT 100x by David J. Eicher

FIND THE OWL NEBULA by aiming your finder scope at the bright star Beta Ursae Majoris and then move the scope 2° southeast, past galaxy M108, to the Owl.

NGC 3132
Magnitude 8.2   Size >47"   by Jack Marling

NGC 3242
Magnitude 8.6   Size 16"   by Jack Marling

NGC 4361
Magnitude 10.3   by Martin Germano

NGC 6210
Magnitude 9.3   by Jack Newton

RINGS, SPHERES, AND BLOBS are all shapes commonly displayed by planetary nebulae. Vela's NGC 3132 (top) is an easily visible ring. NGC 3242 in Hydra (center) has a complex, multilayered structure. NGC 4361 in Corvus (bottom left) displays a curious, disorganized appearance. Tiny NGC 6210 in Hercules is so small that it's often mistaken for a star.

north to two 6th-magnitude stars, and sliding 1.5° west.

Whereas M97 is round and has an elusive central star, NGC 3132 is a bright, ring-shaped nebula with a 10th-magnitude central star. Even a 2-inch telescope at high power shows the 8th-magnitude nebula measuring 30" across with its impossible-to-miss central star. With a careful eye and a 4-inch scope you'll see the planetary's ring shape. An 8-inch telescope equipped with a nebula filter shows that the outer edges of the ring, spanning almost 50", are clearly brighter than the ring's inner parts.

Twenty-two degrees north and slightly east of NGC 3132 lies another large planetary, **NGC 3242**. This object is much more accessible to Northern Hemisphere viewers because it lies 22° higher in the sky. To zero in on this object, start at 4th-magnitude Mu (μ) Hydrae and move your scope 2° south. The nebula lies within the same low-power field as an 8th-magnitude star.

NGC 3242 is a complex object consisting of a bright oval disk surrounded by a much larger, fainter halo of gray-green light. Because of its size, color, and slightly flattened shape, early observers dubbed NGC 3242 the "Ghost of Jupiter." A 6-inch scope shows the bright, flattened oval of nebulosity surrounding a dark central hole. Averted vision reveals the low surface brightness halo of gray-green nebulosity. A 12-inch scope with a nebula filter unveils the 12th-magnitude central star and shows several overlapping layers of increasing brightness, which gives the nebula an almost three-dimensional appearance.

The curious planetary **NGC 4361** lies inside the trapezoid of bright stars that forms Corvus. This nebula shows an unusual bright disk surrounded by an irregularly shaped halo that appears as if it were twisted. Large scopes reveal this faint, curious feature.

To find NGC 4361, aim your scope at Gamma (γ) Corvi (the northwest star in the trapezoid) and swing it southeast by 2.5°. You'll see a pair of 7th-magnitude stars oriented north-south; the nebula lies 30' northeast of the northern star.

The most obvious feature of NGC 4361 is its large size. If you're in the right place, you'll see a patch of fuzzy light. A 6-inch scope shows a 25"-diameter oval-shaped patch of light with two faint wisps on opposite sides extending to 45". Larger telescopes reveal the nebula's 13th-magnitude central star and show a relatively strong greenish color in the nebula's brightest parts, all set in a field sparsely splattered with stars.

The bright green nebula **NGC 6210** in Hercules epitomizes every tiny, difficult-to-identify planetary nebula. Located 3° northeast of Beta Herculis, tucked away beside two stars of equal brightness, NGC 6210 measures only 14" across. Unlike the Owl Nebula (and the other nebulae we've surveyed), NGC 6210 is identifiable as a planetary only because it is green and appears slightly fuzzier than the stars that surround it.

If we could travel to the Owl Nebula, the Ghost of Jupiter, and NGC 6210, we would find that nearly all are the same size. Yet because they lie at different distances, their telescopic appearances vary dramatically — from large and faint M97 to small and bright NGC 6210. Backyard telescopes and patient eyes also show that these objects vary considerably in structure, from rings to disks to complex, overlapping shells. You'll find planetary nebulae to be some of the most varied, unusual forms in the Galaxy. ☐

The Ghostly Glow of Gaseous Nebulae

You'll need to use a variety of observing tricks
and filters to glimpse these extremely faint objects.
by David Higgins

As chilly nights and falling leaves lead us down the road to winter, make sure you get outdoors to see the winter Milky Way, which is visible in the early morning hours. This magnificent glowing band of light contains some of the most spectacular objects in the sky; however most of these objects are challenging targets for backyard telescopes because of their low surface brightnesses. This is especially true of a handful of softly glowing nebulae — winter's ghostly clouds of gas.

One of the most complex and beautiful winter emission nebulae is the **Rosette Nebula** in Monoceros (NGC 2237-9). Measuring 80' by 60', the Rosette Nebula is faintly visible as a soft glow in large binoculars or wide-field telescopes. A wreath-shaped complex of nebulous light, the Rosette contains a distinctive, dark "hole" in its center. Embedded within the nebula is the bright open star cluster NGC 2244, faintly visible to the naked eye from a dark sky. The cluster spans 40' and is rectangular. The bright stars in NGC 2244 are very hot, and their ultraviolet light causes the nebula to shine. A few arcminutes west of NGC 2244 you'll see NGC 2239, a clump of stars associated with the Rosette nebulosity.

Finding the location of the Rosette Nebula is easy, but viewing the nebula is difficult under anything less than a dark sky. To find it, start by locating Epsilon (ε) Monocerotis, a 4th-magnitude star some 7° southeast of Betelgeuse. From this star, move 2° east-northeast, and you'll come upon the star cluster NGC 2244. This group appears as a splash of bright, blue-white stars in finder scopes. The brightest star in NGC 2244 is the magnitude 5.9 yellow giant 12 Monocerotis, probably a foreground star.

Seeing the Rosette nebulosity is somewhat trickier than locating the nebula. The northern portion, NGC 2246, is the brightest part and shows the most detail. NGC 2246's edges are well defined and easily visible, making this part of the Rosette detailed in a 6-inch or 8-inch scope. In larger telescopes dark nebulae are visible at high powers, forming chains and tunnels of dark nebulosity throughout the glowing bright nebulosity.

Low surface brightness nebulae like the Rosette are difficult objects to observe. However, several observing techniques can make them much easier to spot. First, take along a black cloth hood to cover your head and the viewing end of the telescope. By doing this, you'll eliminate scattered light from interfering with your view. Also, use a low-power eyepiece that gives you a wide field of view. Make sure you let your eyes dark adapt for

NGC 2264 Size 60' by 30' by Jack Newton

ELUSIVE NGC 2264 surrounds the bright star S Monocerotis, at right in the photo. The ruddy nebulosity at left outlines a dark cloud of dust dubbed the Cone Nebula.

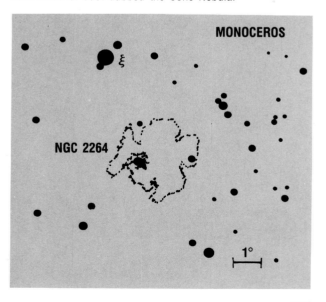

Rosette Nebula Size 80' by 60'
by Tony Hallas and Daphne Mount

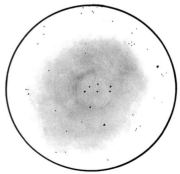

17.5-inch f/4.5 Reflector at 71x
Sketch by David J. Eicher

FILAMENTS OF GLOWING GAS make up the Rosette Nebula in Monoceros.

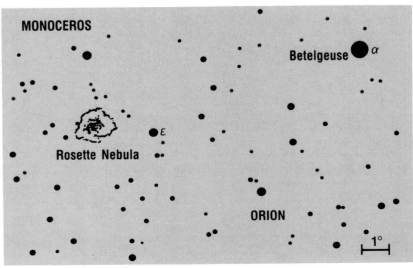

half an hour or so in total darkness before attempting to spot faint details. Finally, use a nebular filter like the Ultra High Contrast (UHC) filter. This cuts out light pollution and lets only the light from the nebula itself into the telescope.

After carefully viewing the Rosette, move your scope 4.5° north and 2° east and you'll find another complex of nebulosity and associated stars, designated **NGC 2264**. The star cluster contains 120 stars and measures some 20' in diameter. Because of its distinctive shape, this object is sometimes called the Christmas Tree Cluster. Large areas of faint nebulosity surround the cluster and cover an area four times that of the Full Moon. On dark nights observers with 10 inch or larger instruments can spot this softly glowing nebula as a large, hazy patch of light.

The most famous object associated with this complex of nebulosity is the Cone Nebula. Although it is a dark nebula set against a very low surface brightness strip of bright nebulosity, the Cone can be glimpsed with small telescopes on extremely dark nights. A UHC filter, averted vision (glancing to the side of the field to use your eye's faint light receptors), and a black cloth hood thrown over your head help in spotting this tough object.

NGC 1931 in the constellation Auriga is a challenging target for scopes in the 6-inch or 8-inch range. Lying a little over a degree west of the open cluster M36, this nebula measures 3' by 3' and glows at about 11th magnitude. Contrary to the nebula's published dimensions, large telescope views show this object as slightly elongated east-west. At high magnifications this elongation suggests a large galaxy with a very bright nucleus.

Another cluster/nebula combination in Auriga is IC 410. A faint, challenging cloud of nebulous light, IC 410 surrounds the bright star cluster NGC 1893. Locate the cluster first by sweeping 2° southwest of the bright star Phi Aurigae. NGC 1893 is a sparse cluster containing some twenty stars spread over 12'.

With averted vision on a dark night, you'll see some of the IC 410 nebulosity with a 6-inch scope. Because it is faint and measures roughly 40' across, IC 410 has a low surface brightness. Consequently, low-power views are the most satisfying. On dark nights 6-inch telescopes show IC 410 as a faint glow with two bright areas aligned north-south; the southern section appears brightest. A large telescope in the 16-inch class reveals that the nebula's southern edge is mottled with dark areas that are speckled across faintly luminous patches.

Enjoy hunting these challenging nebulae this autumn and winter. As the chill pulls at you to go inside, you'll get a great feeling from knowing that these associated nebulae and star clusters are stellar nurseries lodged in the Galaxy's arms. By observing these objects, you're peeking into a region where amazing forces turn old stars into new ones. For me, observing in the cold air is inspiring. After all, the universe shares its secrets with anyone who is willing to approach it. □

These objects are best visible in the winter evening sky.

Challenging Nebulae of the Winter Sky

Object	R.A. (2000.0) Dec.		Size	Tirion	Uranometria
IC 410	5h22.6m	+33°31'	40' by 30'	5	I: 97
NGC 1931	5h31.4m	+34°15'	3' by 3'	5	I: 97
NGC 2237-9	6h32.3m	+05°03'	80' by 60'	11, 12	I: 227
NGC 2264	6h40.9m	+09°54'	60' by 30'	11, 12	I: 182

Object: Designation in NGC or IC catalog. **R.A. and Dec.**: Right ascension and declination in eqinox 2000.0. **Size**: Diameter of nebula in arcminutes. **Tirion**: Chart on which objects appears in *Sky Atlas 2000.0* by Wil Tirion. **Uranometria**: Chart on which object appears in *Uranometria 2000.0* by Wil Tirion, Barry Rappaport, and George Lovi.

IC 410   Size 40' by 30' by Mace Hooley

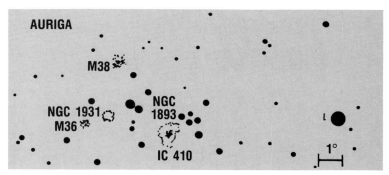

CLUSTER AND NEBULA mingle in the case of NGC 1893 and IC 410 in Auriga.

TINY NGC 1931 is visible at high power as a nebulous spot only 2° northeast of IC 410.

NGC 1931 Size 3' by 3' by Martin C. Germano

The Challenge of Dusty Dark Nebulae

The dust clouds lying in front of bright objects are
among the most exotic and most challenging sights for deep-sky observers.
by David Higgins

Galaxies are full of dust and our Galaxy is no exception. You can see some of this dust by standing outside some night this summer and gazing up at the Milky Way's glowing band of light. When you view the Milky Way from a dark sky, you'll see small lines and clumps of darkness winding through its graceful arch. Although they appear like holes in the Milky Way, these dark patches are actually dust clouds lying in front of the Galaxy's plane. These objects, called dark nebulae, are clouds made of minute, dark dust grains about the size of those in cigarette smoke. Rather than producing or reflecting starlight like bright nebulae, countless particles in each dark nebula absorb light, blocking our view of bright objects that lie beyond the nebulae.

Dark nebulae run the gamut from large and dense to small and tenuous; some are easily observed and others quite elusive. The visibility of a dark nebula depends on its opacity and the density of stars surrounding it. On a dark summer night, several dark nebulae are easily visible in large telescopes and provide a challenge for small scopes. These objects provide a stunning contrast to bright nebulae.

The most stunning dark nebula in small telescopes is **Barnard 92**, named after Edward Emerson Barnard, an astronomer who discovered and catalogued 370 dark nebulae around the turn of the century. B92 is easy to find because it floats in front of the bright, rich starcloud M24 in northern Sagittarius. To find B92, start by locating M24, which is visible to the naked eye as a small, misty cloud of light. Aim your telescope at M24 and sweep around it for a moment, soaking up the great numbers of faint stars. Now move your telescope to the northern edge of the cloud. Using a low-power eyepiece, you'll see two small dark nebulae. The westernmost and larger of the two is B92.

In a 6-inch telescope you'll recognize B92 as a crescent-shaped area devoid of the stars that surround it. Low-power views will show a rich array of stars ringing the outer edge of the field, nicely framing the stark nebula. This aptly demonstrates why early astronomers believed dark nebulae were holes or voids in the Galaxy. A 12th-magnitude star is visible in the nebula and has the appearance of a solitary star basking in empty space. In reality, however, the star is a foreground object.

A 12-inch scope at moderate power shows B92 as a dark patch measuring 12' by 6' with its long axis

B86
Size 4'   by Paul Roques

AN INKY BLACK VOID surrounded by stars, B86 lies several arcminutes west of the star cluster NGC 6520.

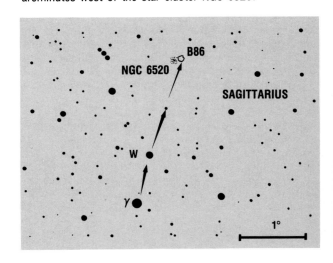

DARK NEBULAE B92 and B93 block light from the bright star cloud M24 in star-rich northern Sagittarius.

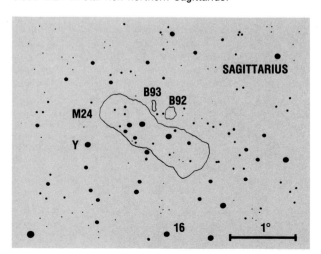

aligned approximately north-south. The nebula's shape is basically oval, but with averted vision two small extensions are visible on its southern end and one is visible on its northern end. The extensions loop into the surrounding starfield, covering 2' or 3' before being swallowed up by stars.

The central star is easily visible with direct vision, and the nebula is bordered by a magnitude 9.4 star on its southern tip and a magnitude 8.4 star on its eastern edge. With averted vision a few faint specks in the nebula flash in and out, eluding direct visibility.

Large backyard telescopes show B92 more prominently because they show many fainter stars around the nebula's edge. With a 17.5-inch scope at high magnification, B92 has a "feathery" appearance because very faint stars are visible in jagged patterns around its edges. This makes it appear like a three-dimensional, black cloud floating in space. The nebula's eastern edge is sharply defined, while its western edge is not as well defined and gradually fades into the stellar background. A 17.5-inch scope also reveals five extremely faint stars within the nebula, in addition to the 12th-magnitude central star.

If you observe B92 with low power, you may notice another dark nebula in the same field to the northeast. Scan 20' northeast of B92 with high power to locate **B93,** which measures 12' x 2'. Although it isn't as opaque or sharply-defined as B92, B93 is also an easy target. B93 is comet-shaped; it has a circular head on its northern end and a tail that flows south over 15'. About 7' south of the head the tail splits in two and forms two distinct dark areas. The northern half of B93 is easily visible in small scopes, but the double tail is difficult or even invisible unless the night is exceptionally dark.

One of the keys to observing dark nebulae like B93 is to slowly sweep across the field using as wide a field as possible. Proper magnification is a must. Large nebulae and nebulae with low opacity both require low magnification to stand out from the stellar background. Small nebulae and high-opacity nebulae can withstand high magnification for close inspection; in fact some appear much better at high powers.

High contrast is also important for observing dark nebulae. You can maximize contrast in two ways. Always observe dark nebulae under a very dark sky. Reduce scattered light in your telescope by baffling the tube, painting the inside of the tube flat black, and keeping the mirror clean.

The star chart you use is equally as important as your telescope. Most modern atlases do not show dark nebulae, but two atlases that do are Antonin Becvar's *Atlas of the Heavens* and volume II of *Uranometria 2000.0*, which covers the southern sky.

The summer sky holds other challenging examples of dark nebulae, some of which are more easily visible than others. Another object in Sagittarius, **B86,** is relatively easy to find because it lies 4° south of the Lagoon Nebula and 2.5° north of Gamma Sagittarii. B86 measures 4' across, is wedge-shaped, and stands out well from the exceedingly rich Milky Way background. The bright open cluster NGC 6520 lies just 5' east of B86. The two objects are visible together in practically any field of view, even at moderately high magnification.

Small scopes show B86 as a square void immediately south of a magnitude-6.9, orange star. Another orange, 7th-magnitude star lies on the nebula's north-

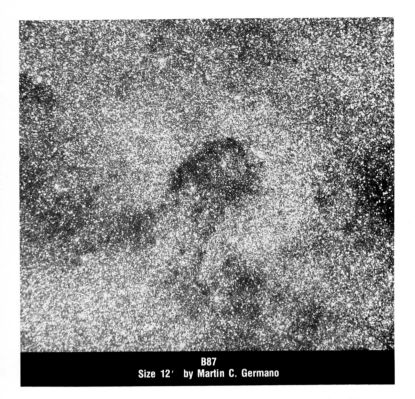

B87
Size 12' by Martin C. Germano

A BIRD IN PROFILE, elusive B87 (above) earns its nickname, the Parrot's Head Nebula. Barnard's Snake Nebula (below) carves an S-shaped hole in the sky.

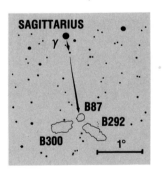
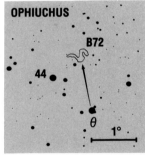

B72
Size 4' by Paul Roques

Dusty Dark Nebulae

Object	R.A. (2000.0) Dec.		Size	Uranometria
B72	17h23.5m	−23°38'	4'	II: 337, 338
B86	18h02.7m	−27°50'	4'	II: 339
B87	18h04.3m	−32°30'	12'	II: 377
B92	18h15.5m	−18°11'	12' by 6'	II: 339
B93	18h16.9m	−18°04'	12' by 2'	II: 339

Object: Designation in Barnard catalog. **R.A. and Dec.:** Right ascension and declination in equinox 2000.0. **Size:** Diameter in arcminutes. **Uranometria:** Chart on which object appears in *Uranometria 2000.0* by Wil Tirion, Barry Rappaport, and George Lovi.

western edge. Unlike B92, B86 has edges that gradually fade into the surrounding background of stars. The nebula's high opacity alone makes it stand out well.

Another dark nebula in southern Sagittarius is **B87**. This round nebula lies 2° south and a few arcminutes west of Gamma Sagittarii. B87 is far more challenging than B92 for small telescopes because its center is filled with myriad stars. The result is a circular dark nebula 12' across with a low opacity. B87 has a curved extension to the east, and its eastern side appears slightly darker than its western side. Two larger, less opaque dark nebulae, B292 and B300, lie southwest and southeast of B87. To Barnard and other early observers, the three nebulae together resembled a winged bird. Because of this, B87 is nicknamed the Parrot's Head Nebula. An 8th-magnitude star lies just east of the nebula's center, marking the bird's eye.

One of Barnard's most famous nebulae is **B72**, an elusive object located in Ophiuchus. Because of its S-shape, B72 has been given the nicknames "S Nebula" and "Snake Nebula." Its unique shape makes B72 a sought-after object for amateur observers. To find B72, aim your telescope at Theta Ophiuchi and sweep 1.5° north-northeast. (The nebula can also be found by moving less than a degree northwest of 44 Ophiuchi.) If the sky is exceptionally dark, a low-power field will show a thin, winding dark lane that covers nearly 30' and is oriented east-west. The width of the winding dark lane varies between 2' and 3' throughout its path. With averted vision you should see that the dark lane forms an "S."

Observing dark nebulae will challenge you and your telescope on different levels. B92, B93, and B86 are high-contrast nebulae so viewing them under moderately dark skies is straightforward. On particularly dark nights, try to resolve and count individual stars within these nebulae to test the acuity of your eye. The lower-opacity nebulae, B87 and B72, are much more difficult to observe. Simply seeing them or tracing their shapes is a challenge. Whichever challenge you take, you're certain to discover an entirely new type of observing when you aim your scope at the dusty blotches called dark nebulae. □

These objects are best visible in the summer evening sky.

Tracking Down the Helix

It is king of the planetary nebulae: largest, brightest, and closest. It is also one of the most challenging to observe with a small telescope.

by David Higgins

Planetary nebulae are beloved deep-sky objects. Compact, bright, and colorful, they show details in amateur telescopes both large and small. Their characteristic high surface brightnesses allow observers to see details like rings, central stars, and brightness variations in disks of light from across the Galaxy. Yet not all planetaries are so easy to observe, and this is because of their low surface brightness. Ironically, this includes the king of planetary nebulae, an object that despite its great size and brightness is one of the most challenging objects of all.

The Helical Nebula (NGC 7293), so named because its shape resembles that of a double helix, is the closest planetary nebula. (The nebula is also informally called the Helix.) Because it lies only about 450 light-years away, it is also the largest and brightest planetary in the sky. How can such a close, bright, and large object be challenging to view in a small telescope? Surprisingly, its large size is the main culprit. Overall the Helical Nebula equals a magnitude 6.5 star in light output. This total magnitude is a measure of the object's brightness if it were squeezed into a point source. But the Helix measures about 13' across, so its light has a low surface brightness. In other words, it is spread out so much that little parts of it appear very dim. This is the irony of the Helix Nebula and many other deep-sky objects that are large and nebulous.

Finding the position of the Helix is easy to do on these crisp fall evenings. Start by aiming your finder scope toward the 3rd-magnitude star Delta Aquarii. Extend a line south to Fomalhaut, the 1st-magnitude star in the constellation Piscis Austrinus. (Although this faint constellation barely peeks over the horizon for most Northern Hemisphere observers, Fomalhaut is bright enough and high enough to be used for star hopping.) Look about halfway between these two stars and about 5° west and you'll see the 5th-magnitude star Upsilon Aquarii. Next, move 1° west and you'll have captured the Helix in a low-power field. But whether or not you can see the Helix depends on several important factors.

The first consideration is the instrument. You can use a pair of large binoculars that offer good contrast and sharp images, but be prepared to steady them with a tripod or by bracing them against your knees. A somewhat better instrument is a rich-field telescope, a short-focus instrument (f/3.5 to f/6) that provides a wide field of view. If you use a normal telescope, be prepared to insert a low-power eyepiece that yields at least a 1° field of view. The idea is to provide yourself with a comfortable amount of dark sky surrounding the nebula so you can see it against the background sky.

Because the Helix culminates low in the sky for most northern observers, attempt to view it only when it lies near the meridian — when it's at the highest possible point in the sky from your viewing location. This minimizes the amount of image-degrading atmosphere you need to look through to see the nebula's image.

A good nebula filter helps tremendously with low-contrast objects like the Helical Nebula. Place it on your low-power eyepiece before attaching it to the telescope and when you arrive at the correct position slowly sweep back and forth in the area. The movement of the ghostly light from the Helix will help your eye detect its presence.

Once you locate the Helical Nebula, what you'll

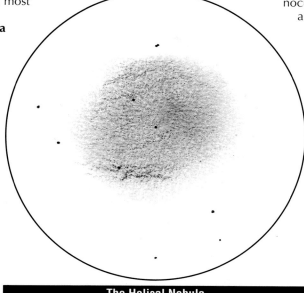

The Helical Nebula
6-inch f/8 reflector at 70x   Sketch by Chesley Runyon, Jr.

The Helical Nebula
Mag. 6.5   Size 13'   by Jack Newton

see depends on your telescope and the sky conditions. On exceptionally dark nights, a pair of 7x35 binoculars reveals a faintly glowing donut of light. Small telescopes show a faintly glowing ring of light some 10' across, with a slightly darker center and a 13th-magnitude central star. Larger telescopes will show some detail in the Helix. A 10-inch instrument shows the nebulosity much brighter and the central hole much darker than a 6-inch scope shows. The 10-incher also reveals a sprinkling of stars throughout the nebulosity and several arcs around the edges that appear brighter than most of the rest of the nebula.

In a large backyard scope the view changes considerably with the added light-gathering power. A 17.5-inch scope at 100x provides a memorable view of the Helix. The nebulosity fills the eyepiece field. The southern and eastern edges are visible as the nebula's brightest portions, and the western half contains several faint stars. I get the impression that the nebulosity is very faintly extended north-south, but even with a 17.5-incher I've never been able to directly see the helical coil shape.

The best way to maximize the detail you see is with a specific type of nebula filter. Try a UHC or O III filter and you'll be surprised at just how much you will see. Using the UHC filter with a large backyard scope, the Helix is elongated north and south and appears faintly helical in shape. The central star appears much dimmer, but the central hole is really obvious. The eastern and southern edges are still the brightest, even with the filter. The most obvious feature using the O III filter is the appearance of a faint milky nebulosity within the central hole. The extension of the nebulosity north and south is not visible with the O III so the helical coil is not visible.

After you've finished exploring the Helical Nebula, it's time to turn your gaze toward another planetary nebula in Aquarius. **The Saturn Nebula** (NGC 7009) measures 25" across and shines at 8th magnitude, making it an easy target for small telescope observers. Although this object was discovered in 1782

The Saturn Nebula
Mag. 8.3P   Size 25"   by Dan Marcus and Bill Smith

M2
Mag. 6.5   Size 12.9'   by Bill Iburg

Deep-Sky Objects in Aquarius

Object	R.A. (2000.0)	Dec.	Type	Mag.	Size
M73	20h59.0m	-12°38'	OC	8.9$_P$	2.8'
NGC 7009	21h04.2m	-11°22'	PN	8.3$_P$	25"
M2	21h33.5m	-0°49'	GC	6.5	12.9'
NGC 7293	22h29.6m	-20°48'	PN	6.5$_P$	13'

Object: Designation in Messier or NGC catalog. **R.A. and Dec.:** Right ascension and declination in equinox 2000.0. **Type:** GC = globular star cluster, OC = open star cluster, PN = planetary nebula. **Mag.:** V magnitude except subscript P, which denotes photographic magnitude. **Size:** Diameter in arcminutes or arcseconds.

by William Herschel, NGC 7009 was nicknamed the Saturn Nebula half a century later by William Parsons, the Third Earl of Rosse. Parsons noticed what looked like two extensions, or ansae, extending from the ends of the planetary. Because the ansae resembled the rings of Saturn, Parsons created a nickname that stuck.

To find NGC 7009, start by locating the magnitude 4.5 star Nu Aquarii. Simply move 1.5° west and you'll come upon NGC 7009. Because this object is so small (in marked contrast to the Helix), it will appear like a tiny, bright blue disk. A 6-inch scope shows the Saturn Nebula as a pale, blue-green oval that is somewhat elongated east-west. In telescopes 10 inches or more in aperture, the nebula's center is brighter than the disk and the thin, faint extensions are visible on the eastern and western sides.

In large backyard scopes the Saturn Nebula appears appreciably more blue. A 17.5-inch instrument easily shows the nebula's ansae. At high magnifications, the haze surrounding the bright core is somewhat darker on the east side. Try a UHC filter and see what a difference it makes. An O III filter, on the other hand, does not seem to improve the view.

Now move about 2° southwest of NGC 7009 and you'll visit an intriguing Aquarian deep-sky object. **M73** (NGC 6994) is an open cluster that was discovered in 1780 by Charles Messier. This cluster, considered by some astronomers to be only a chance alignment of stars, is quite sparse. In fact it consists of four faint stars arranged in a Y-shaped group. Measuring 2.8' across and glowing at photographic magnitude 8.9, this cluster certainly holds the distinction as the plainest Messier object. Nevertheless, it's a fun object to observe simply so you can later say you've seen it. Messier may have believed he observed nebulosity involved with these stars — something he inferred in his notes — and included the object in his catalog because of it. But all modern studies show only four stars in good old M73, so it remains as one of the great curiosities in deep-sky observing.

If you're a globular cluster fan, then Aquarius does offer another that will provide you with a most impressive view. **M2** (NGC 7089) is another of Messier's objects, but much larger and brighter than M72. Measuring 12.9' across and shining at magnitude 6.5, M2 is easily visible as a round fuzzy ball in most finder scopes. The easiest way to find M2 is by starting with the 3rd-magnitude double star Beta Aquarii. Extend a line about 5° north of Beta and you should have M2 in your finder scope. In 4-inch or smaller scopes, M2 gives the appearance of a fuzzy ball with a sprinkling of stars around the edges.

In 8- to 10-inch scopes, the edges are resolved while a number of stars become visible across the face of the cluster. In medium-sized scopes, the core is brighter than the surrounding haze of the cluster. In large backyard scopes, M2 is an awe-inspiring sight. In large telescopes, beyond the 17.5-inch range, high magnifications show the core as resolved and granular. Although not as large or quite as bright as M13 or M22, M2 is nevertheless a heavyweight among globular clusters and fascinating to observe.

Whether you are fascinated with deep-sky challenges like spotting the Helix, resolving distant globular clusters, or tracking down the star-group M73, Aquarius offers something for you. Couple this with the moderate weather that comes with autumn, and you'll have a combination that can't be beat. □

These objects are best visible in the autumn evening sky.

The Horsehead Nebula
Size 6' by 4'   by Jack Marling

Stalking the Elusive Horsehead

With a moderate-sized telescope, a black sky, and a little patience,
you can find one of the sky's most beautiful deep-sky objects.
by David Higgins

Orion's legendary Horsehead Nebula represents the greatest challenge of all for deep-sky observers. Unrivaled as a tough object for visual observers, the Horsehead is far more elusive than photographs suggest. Although relatively easy to capture on film, this small, dark nebula consistently evades those who wish to see it with eyes alone.

The Horsehead is a cloud of dust particles suspended in space just south of Orion's belt. Dark nebulae like the Horsehead are strewn throughout the spiral arms of the Milky Way. Because they don't emit light, dark nebulae are visible only when they block our view of something bright behind them. In the case of the Horsehead, the nebula lies directly in front of a dim strip of emission nebulosity catalogued as IC 434. The Horsehead, itself catalogued as Barnard 33, is visible as a tiny, dark notch in the eastern edge of IC 434.

The Horsehead picked up its nickname in the nineteenth century because, well, it *is* shaped like a horse's head. Ever since its discovery, the allure of seeing such a challenging object with a curious and distinctive shape has drawn thousands of backyard astronomers to the eyepiece in search of the Horsehead. Observer's reports have proven that the Horsehead can be seen with a telescope as small as 5-inches in aperture, although more typically it remains a challenge for those armed with 12-inch or even 16-inch telescopes. Key factors in spotting the Horsehead are observing under a perfectly black sky, pinpointing the exact location of the nebula, achieving maximum eye sensitivity through dark adaptation, and viewing with a nebula filter.

Orion — Here I Come!

Before heading outside to observe the Horsehead, study the region of central Orion carefully using the map on page 84. IC 434 lies embedded in a large area of hydrogen gas and dust that stretches from Alnitak (Zeta [ζ] Orionis) south to Rigel. Long-exposure photographs show that this area is very rich in gas and dust. However, most of this nebulosity — with the exception of the Orion Nebula and the Horsehead region — is beyond visual detection.

Orion's belt is composed of three stars: Alnitak, Alnilam (Epsilon [ε] Orionis), and Mintaka (Delta [δ] Orionis). Alnitak, the easternmost star, is surrounded by nebulae. IC 434 is aligned north-south, begins at Alnitak, and extends 1° to the south. The Horsehead Nebula lies about 40' south of Alnitak. The Horsehead measures 6' by 4' and is the darkest part of a cloud of dark nebulosity that borders IC 434. Other nebulae lie in the immediate area: for example, NGC 2024 is a large circular nebula immediately east of Alnitak that is bisected by a broad lane of dark nebulosity. This object serves as a good test before you try for the Horsehead. NGC 2024 is far brighter than IC 434; so if NGC 2024 is easily visible, conditions are favorable for spotting the Horsehead. Another test object is the small, roundish reflection nebula NGC 2023, located about 25' south of Alnitak. This nebula surrounds an 8th-magnitude star and is far easier to see than its neighbor IC 434.

The simplest way to find the Horsehead is to start by zeroing in on NGC 2023. With a 6-inch or larger scope and a medium-power eyepiece, aim your scope 25' southeast of Alnitak and you will see a slight haze surrounding an 8th-magnitude star. (Make sure that you're using a high enough magnification to keep Alnitak outside of the field of view. Otherwise, the bright

What Is the Horsehead, Anyway?

The Horsehead Nebula is a vast cloud of gas intermingled with dark dust particles; the particles are roughly the size of those in cigarette smoke. Most dark nebulae contain predominantly hydrogen gas and are typically 10 or 20 percent dust.

Most dark nebulae are not so easily visible as the Horsehead. For astronomers to detect a dark cloud, it must lie along a line of sight that places it directly in front of a reasonably dense background of bright material, like a rich starfield or a bright nebula. The Horsehead is visible only because it blocks light from the emission nebula IC 434.

Although it's easy to think of celestial objects as static, particles in nebulae are in a constant state of motion. Only the Horsehead's large physical size and great distance make it seem changeless. But change will come given time: Over the course of several thousand years, the object we know as a Horsehead may very well transform itself into a not-so-glamorous mule.

star's glare will interfere with seeing NGC 2023.) After identifying NGC 2023, fix your gaze approximately 10' west of the nebula's central star and you'll see a lone, 9th-magnitude star. Now look about 5' south of this star; you'll be staring directly at the position of the Horsehead Nebula.

Hints for Horsehead Hunting

If at first you don't see the Horsehead or at least the ghostly glow of IC 434, you can employ a few tricks of the trade to heighten your chances of success. First, give your eyes ample time to fully dark adapt to their maximum sensitivity. This generally means "warming up" with some casual observing for a period of half an hour or more.

To provide the best possible dark adaptation, try to keep your exposure to the Sun at a minimum for a day or two prior to observing. Additionally, use averted vision when you're searching for the Horsehead. A gradual glance off to the side of the telescope's field lets you see an object in the field center with your rods, your eye's specialized faint-light receptors. This can make the difference between seeing an extremely faint object and not seeing it.

Make sure that your observing site is very dark and that you attempt to observe the Horsehead on nights of excellent seeing and transparency. Also, your best Horsehead observing will be possible when central Orion is on or near the meridian, at its highest possible elevation from your site. At such a time you'll be looking through the least possible amount of particles in Earth's atmosphere, minimizing the interference with the precious few photons traveling all the way from IC 434.

PINPOINTING THE HORSEHEAD is no easy task. Start at Alnitak, the easternmost star in Orion's belt. From there, move 30' south: the Horsehead lies 5' east of a faint pair of stars. Photo by Greg Field. Map copyright 1987 Willmann-Bell, Inc.

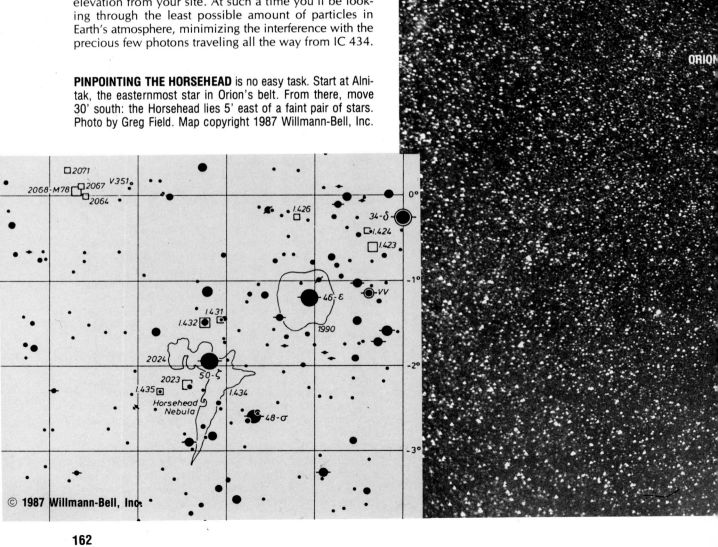

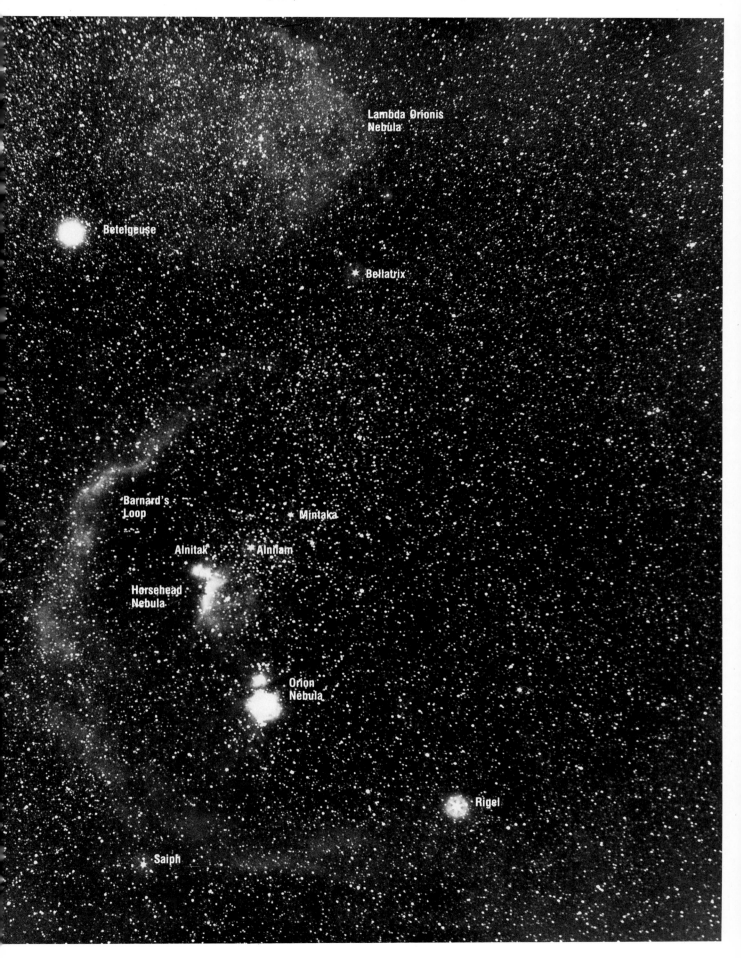

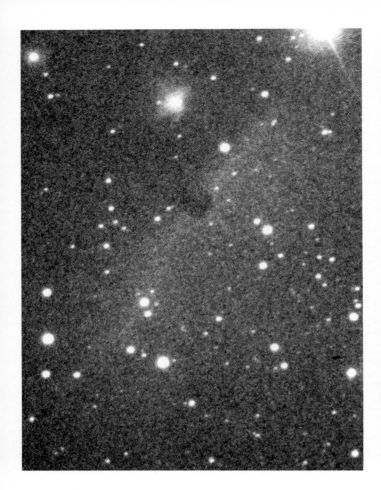 

SIMULATED VIEWS of the Horsehead Nebula dramatize the effect of large-aperture viewing. A 10-inch telescope barely detects the Horsehead (left), whereas an 18-inch provides a reasonably detailed view (right). Photos by Hal Jandorf (left) and Jack Newton.

To better the odds of seeing the Horsehead, use a nebula filter. Specifically, the Lumicon H-β filter is designed to boost contrast on faint nebulae like IC 434, giving observers with small telescopes the best possible chance to see a faint object. Under ideal conditions the Horsehead is visible without a filter, but using the H-β filter will increase your chances of success.

Seeing the Horsehead

After you recognize the faint glow of IC 434 and pinpoint the Horsehead, the level of detail you'll see depends on your telescope, observing experience, and the quality of the night. Telescopes in the 6-inch to 10-inch range show an extremely faint wisp of nebulosity, visible only by averted vision, less than 30' in length. Under average conditions the faintness of IC 434 may make you wonder if you can see it at all.

The Horsehead will most likely appear as an indistinct nick 1' long cut into the central, eastern portion of the band of nebulosity. Large telescopes make seeing the Horsehead much easier. A 14-inch or 16-inch telescope shows IC 434 as a faint strip of nebulosity nearly 60' long. The Horsehead appears like a small, square box, without the familiar features — nose and mane — that stand out in photographs. Only in extremely large backyard scopes — 20-inchers or larger — will some hint of the horse's nose be visible.

Increased contrast from using the H-β filter greatly improves the view. With small telescopes it yields a view of IC 434 that leaves you certain you actually saw it. The edges of IC 434 are no longer indistinct: they are well delineated from the surrounding starfield. In telescopes of 16-inches and larger, IC 434 is much more easily visible, which in turn makes the Horsehead stand out without too much trouble on dark nights. With a 17.5-inch scope on a good night, the filter allows you to see the Horsehead plainly with direct vision.

A filtered view through my 24-inch scope shows some shape in the Horsehead; the head component appears rounded on its southwestern side and slopes toward the horse's nose. The southern side, or mane, is the darkest part of the nebula. The horse's neck widens as it fades into the dark cloud that extends eastward. On nights of superb transparency, I can glimpse several faint field stars visible immediately north of the nose.

The final word on Horsehead hunting is experimentation. Reports from experienced and reliable observers establish that in some cases the Horsehead is visible in a 5-inch scope, in others invisible in a 40-inch scope. Try different eyepieces and telescopes to give yourself a wide choice of views of the Horsehead and IC 434. Some eyepieces will show more detail than others. And experiment with filters, dark adaptation, and averted-vision techniques.

On top of everything, keep searching. If you do, you'll see the Horsehead for yourself. And after you see the Horsehead, other deep-sky challenges will no longer seem quite so intimidating. □

These objects are best visible in the winter evening sky.

The Horsehead Nebula
photo by Michael Stecker

Find a Supernova Remnant

Careful searching with a moderate-size telescope
can show the faintly glowing remains of stars that exist no more.
by Chris Schur

To see a supernova remnant is to see one of the most short-lived and tenuous of all deep-sky objects. Cosmic bubbles blown off by dying stars, they balloon into space for tens of thousands of years before becoming so thin that they are no longer visible. While tens of thousands of years is a long time by human standards, it is but a moment in the history of the universe. As deep-sky objects go, observing supernova remnants is an extreme challenge. Their low surface brightnesses mean that even relatively close, young supernova remnants are difficult to spot. And as they get older their surface brightnesses plunge.

Supernova remnants result from the most violent events in the universe, save for the Big Bang itself. When an extremely massive star runs out of material for nuclear fusion, it begins a process that leads to internal collapse and finally an enormously powerful explosion. The energy released during this explosion can equal the total output of one hundred million Suns. The supernova can, briefly, outshine an entire galaxy.

The most easily observed supernova remnant is the **Crab Nebula** in Taurus, catalogued as M1 and NGC 1952. The Crab is the result of a supernova explosion that occurred just 900 years ago. (The event was observed in A. D. 1054 by Native American skywatchers and meticulously recorded by Chinese astronomers.) It lies at a distance of about 6,300 light-years and is only a few light-years in diameter.

The Crab Nebula was discovered telescopically by the English amateur astronomer John Bevis in 1731. All that is left of the once mighty progenitor star is a 16th-magnitude pulsar, rapidly spinning 30 times a second.

To find the Crab Nebula, start by locating the bright star Zeta (ζ) Tauri. Now move slightly over 1° northwest and you'll see the Crab centered in the field.

At about 9th magnitude the Crab is faintly visible in large binoculars as a tiny patch of gray-green light. Small telescopes reveal much more of the nebula's shape, which is more or less oval with some irregular mottling around the edges. An 8-inch telescope at 50x reveals an irregularly shaped, elongated nebulosity measuring about 6' by 4' situated in a rich star field. In a 12-inch scope or larger you may see variations in brightness inside the nebula and also intricate patterns of fine, linear streaks of nebulosity across the face of the object. These are the brightest of the Crab's filaments, glowing spikes of material that condensed after the explosion.

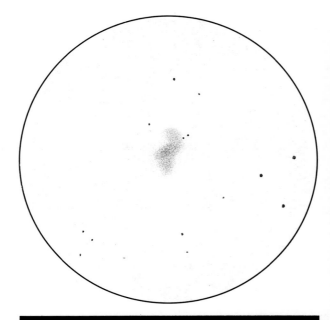

The Crab Nebula
8-inch f/10 SCT at 100x   Sketch by Chris Schur

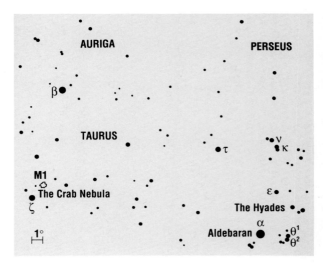

THE CRAB NEBULA is the brightest supernova remnant in the sky. It is located 1.5° northwest of Zeta Tauri.

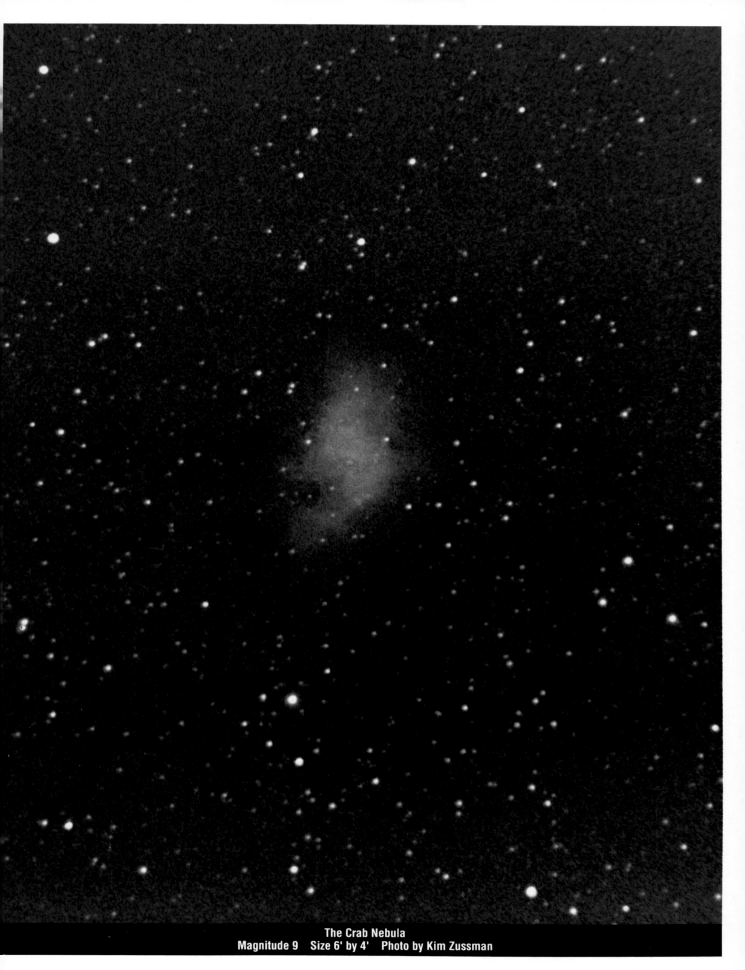

The Crab Nebula
Magnitude 9   Size 6' by 4'   Photo by Kim Zussman

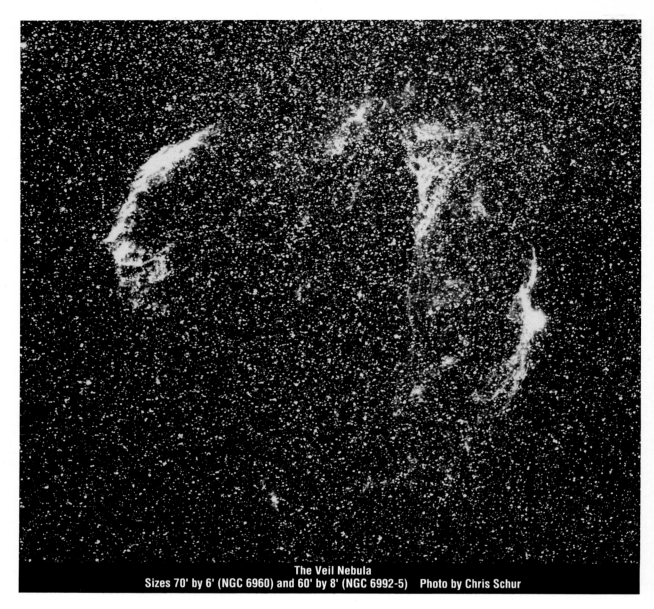

The Veil Nebula
Sizes 70' by 6' (NGC 6960) and 60' by 8' (NGC 6992-5)   Photo by Chris Schur

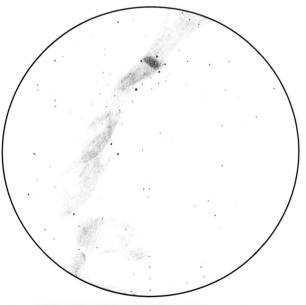

NGC 6992-5
16-inch f/4.4 reflector at 90x   Sketch by Rick Rotramel

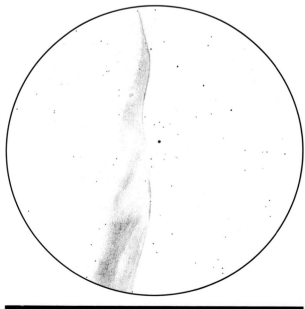

NGC 6960
16-inch f/4.4 reflector at 90x   Sketch by Rick Rotramel

NGC 6888
Size 20' by 10'   Photo by Scott Hicks

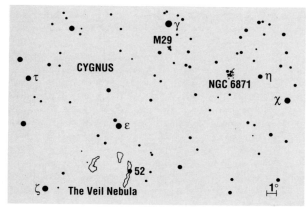

GHOSTLY SHELLS OF LIGHT, the Veil Nebula and NGC 6888 are both challenging targets. Nebula filters enhance their faint glows and make them appear brighter.

Because most supernova remnants cover at least a degree, observing them requires a wide-field eyepiece. The large exit pupils provided by these oculars render bright and contrasty images that enable you to see faint detail. Nebula filters allow you to boost the contrast between an object and the sky background further by transmitting the nebular light and cutting out light pollution. The result is a darker sky background that makes the nebula appear to "pop."

This contrast enhancement is crucial for spotting supernova remnants since much of their light is only slightly brighter than the sky background itself. To maximize your view of faint details, slowly nudge the telescope's tube back and forth as you try to pick out detail in the nebula. This helps bring out details that may be missed in a stationary view because your eyes are especially sensitive to motion when viewing faint light.

An object much improved by field sweeping is the **Veil Nebula** in Cygnus, one of the largest supernova remnants. Discovered in 1784 by William Herschel, the Veil Nebula is all that remains of a supernova explosion that occurred some 30,000 or 40,000 years ago. Located 2.5° south of the bright star Epsilon (ε) Cygni, the Veil spans some 2.6°, which equals 70 light-years at its distance of 1,500 light-years.

The Veil Nebula is one of the most detailed objects in the sky. Consisting of two bright, narrow components and a faint triangular patch, the Veil appears dim because its light is spread over such a large area. A wide-field rich-field telescope under a moonless sky may show the entire Veil Nebula as an extremely faint oval of gray light some 2° in diameter. However, the oval appears incomplete; only the eastern and western edges are clearly defined.

Large telescopes show the individual parts of the Veil Nebula. The easiest segment to find is NGC 6960, a long, thin arc of nebulosity oriented north-south that is bisected by the bright star 52 Cygni. Six-inch telescopes reveal traces of this object, although a 10-incher is required to see the object's knotty detail, which resembles twisted strands of rope glowing with an eerie light. The eastern section, NGC 6992-5, is brighter and considerably more detailed than NGC 6960, but it is more difficult to find.

To locate NGC 6992-5, sweep about 2° due east of 52 Cygni. Both NGC 6960 and NGC 6992-5 measure approximately 65' by 7'. Lying between the two bright sections is a much more challenging object, the triangular nebula NGC 6979. This object is not easily

IC 443
Size 50' by 40'   Photo by Chris Schur

Supernova Remnants for Backyard Telescopes

Object	R.A. (2000.0) Dec.		Mag.	Size
M1 (NGC 1952)	5h34.5m	+22°01'	9	6' by 4'
Simeis 147	5h39.1m	+28°00'	—	200' by 180'
IC 443	6h16.9m	+22°47'	—	50' by 40'
NGC 6888	20h12.0m	+38°21'	—	20' by 10'
NGC 6960	20h45.7m	+30°43'	—	70' by 6'
NGC 6992-5	20h56.4m	+31°43'	—	60' by 8'

Object: Messier, NGC, or other catalog designation. **R.A. and Dec.:** Right ascension and declination in equinox 2000.0. **Mag.:** Approximate V magnitude of nebula. **Size:** angular extent in arcminutes.

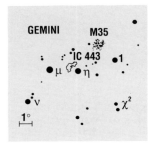

IC 443 IN GEMINI appears as an extremely faint loop of gray light in large telescopes.

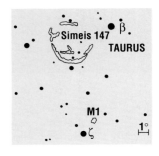

THE ULTIMATE CHALLENGE in photographing supernovae is Simeis 147.

Simeis 147
Size 200' by 180'   Photo by Chris Schur

visible with small telescopes but can be spotted on a dark night with a 10-inch scope.

Also in Cygnus is the more challenging supernova remnant **NGC 6888**, located in an extremely crowded star field about 3° southwest of Gamma (γ) Cygni. This object is an oval nebulosity measuring 18' by 12' and is a challenging test for an 8-inch telescope.

Two progressively more difficult supernova remnants lie in the winter sky. Located just 3° southeast of M35, **IC 443** can be glimpsed in a 10-inch scope equipped with a good nebular filter. At a distance of 10,000 light-years, this object has a much lower surface brightness than the Crab or the Veil. About the same age and size as the Veil, IC 443 is also circular in form. Unlike the Veil, however, IC 443 expanded into a nonuniform interstellar medium and is much brighter on its eastern side.

The eastern portion is the part you can see in backyard scopes. To spot it, sweep about a degree east of the 3rd-magnitude star Eta (η) Geminorum. In a low-power ocular it appears as a 25' by 5' arc of nebulosity, glowing faintly against the sky background. Even the largest scopes show little more, as most of the fine structure in this nebula is either too faint to be seen or smaller than an arcsecond in width.

Some supernova remnants are too faint to see by eye. Here, the power of photography serves to reveal these objects. Located just 5° north of the Crab Nebula is **Simeis 147**. At a distance of about 1,500 light-years, S147 is very close as such objects go. Although comparable to the Veil Nebula in size, it is much fainter and has, as far as I know, never been seen visually. However, it remains a challenge to backyard astrophotographers who wish to capture the extremely faint wisps of light from this tenuous object. I've found that hypered Tech Pan and a deep, filtered exposure successfully record the brightest parts of this extremely difficult object.

Hunting supernova remnants requires patience, persistence, and a very dark sky. When all three factors are working together you will be able to witness the remains of nature's most violent act. That's a mind-boggling thing to think about as you gaze at a fuzzy disk of light in your telescope's eyepiece on peaceful little Earth. □

These objects are best visible in the autumn evening sky.

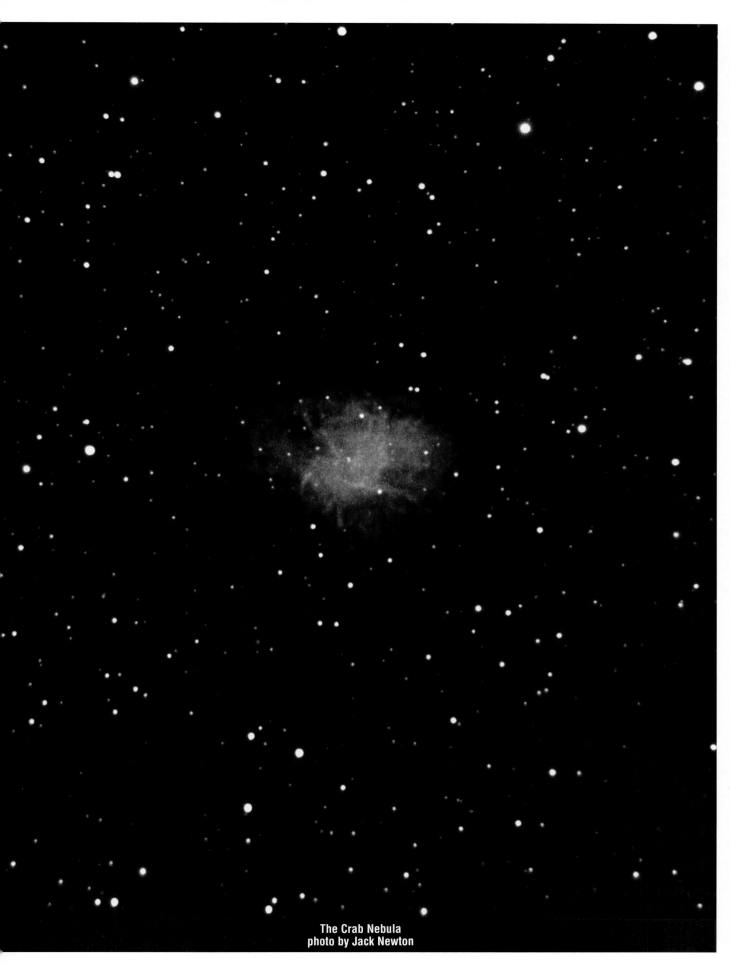
The Crab Nebula
photo by Jack Newton

The Challenge of Winter Nebulae

Push your telescope and your observing skills to the limit by finding these incredibly faint wisps of light.

by Phil Harrington

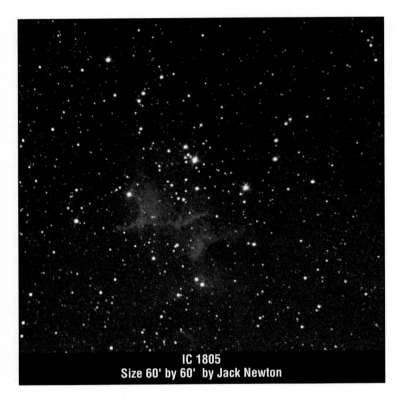

IC 1805
Size 60' by 60' by Jack Newton

Nebulae get a bum rap. Many amateur astronomers think that most nebulae are only visible in long-exposure photographs. While this might have been true twenty or thirty years ago, the popularity of large, fast Newtonian reflectors, advanced multi-coated eyepieces, and contrast-enhancing nebula filters have all helped to bring many once-too-faint nebulae into the realm of backyard astronomers.

In fact several examples of challenging emission and reflection nebulae are scattered across the winter sky. These objects are not easy to spot, mind you, but they are within the grasp of moderate-sized backyard scopes and will challenge the combination of your eyes, observing site, and telescope.

As a warm up, begin with two huge loops of nebulosity in Cassiopeia the Queen, which is known to deep-sky observers as the home of many pretty open clusters. Just beyond the eastern tip of Cassiopeia's W-shaped asterism is the pair of star clusters **IC 1805** and **IC 1848**. Both are visible in binoculars and finderscopes as small, tight knots of three or four stars. Switching to a 6-inch or larger telescope unmasks each as a large, loose open cluster. Each of these star clusters has a large cloud of nebulosity associated with it. The star clusters are easy to see; the nebulae are more challenging.

IC 1805 (labeled as Melotte 15 on some star atlases) is the richer of the two. Within this star group's

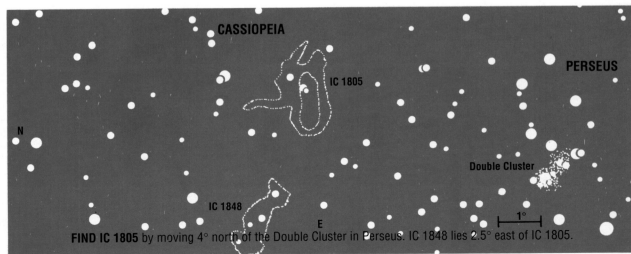

FIND IC 1805 by moving 4° north of the Double Cluster in Perseus. IC 1848 lies 2.5° east of IC 1805.

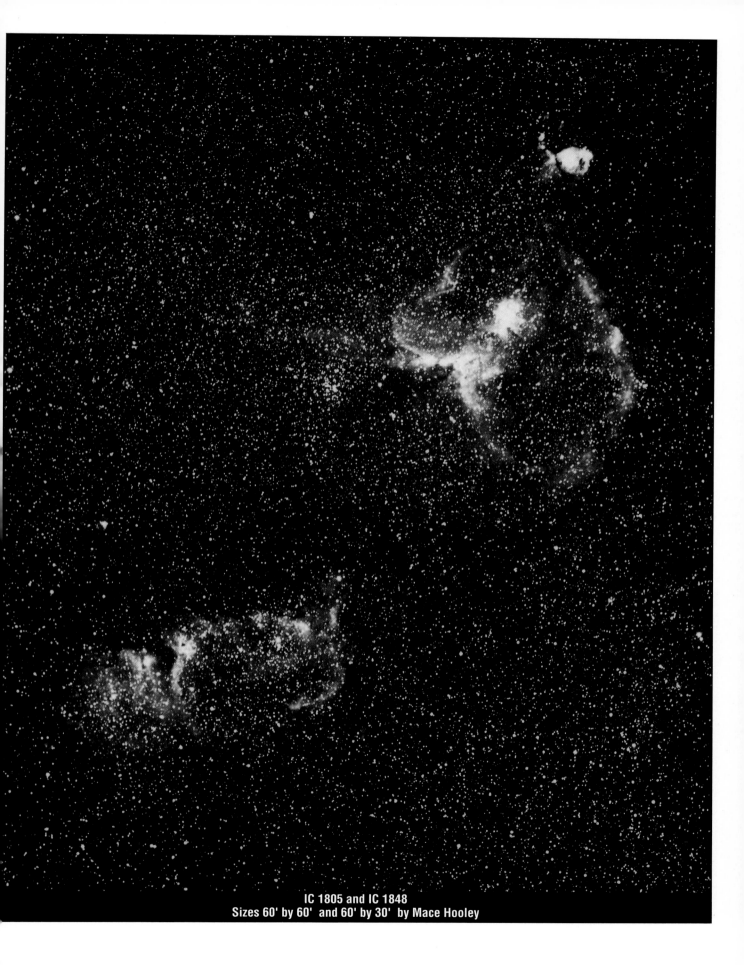

IC 1805 and IC 1848
Sizes 60' by 60' and 60' by 30' by Mace Hooley

IC 59 and IC 63
Sizes 10' by 5' and 10' by 3' by Martin C. Germano

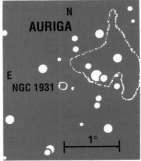

EMISSION NEBULA NGC 1931 is located 1.5° east of the 5th-magnitude star Phi Aurigae.

IC 59 AND IC 63 lie 20' north and east of the bright star Gamma Cassiopeiae.

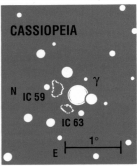

CHALLENGING WINTER NEBULAE

Object	R.A.(2000.0)Dec.	Type	Size
IC 59	0h56.7m +61°04'	E+R	10' by 5'
IC 63	0h59.5m +60°49'	E+R	10' by 3'
IC 1805	2h33.4m +61°26'	E	60' by 60'
IC 1848	2h51.3m +60°25'	E	60' by 30'
NGC 1579	4h30.2m +35°16'	R	12' by 8'
IC 405	5h16.2m +34°16'	E+R	30' by 19'
IC 410	5h22.6m +33°31'	E	40' by 30'
NGC 1931	5h31.4m +34°15'	E+R	3' by 3'

Object: Designation in NGC or IC catalog. **R.A. and Dec.:** Right ascension and declination in equinox 2000.0. **Type:** Emission (E) or reflection (R) nebulosity. **Size:** Diameter in arcminutes.

grasp lie forty 8th-magnitude and fainter stars splashed across a 22' diameter. To find IC 1805, start at Epsilon (ε) Cassiopeiae, the easternmost star in the W-shaped asterism. Sliding just over 3° east, watch for a pair of 7th-magnitude stars set in a rich Milky Way field. From here, continue another 2.3° southeast, where you will spot IC 1805.

Surrounding the cluster's stars are the faint remains of the clouds from which they were formed. The brightest portion of the IC 1805 nebula lies southwest of the cluster's brightest stars. Under an exquisitely dark sky last winter, my 13.1-inch f/4.5 Newtonian revealed faint, grayish arcs of light set between the clusters. Although no fine detail was visible, the cloud's eastern boundary appeared sharp and distinct. The western edge was diffuse and ill-defined.

Because the IC 1805 nebulosity is spread across a full degree of sky, be sure to use your lowest-power eyepiece when searching for it. Even then it may evade you. The nebulosity is difficult to spot for several reasons. First, its huge size gives it a low surface brightness — its light is spread out so that individual parts of it appear dim. This demands the light-gathering ability of a large-aperture telescope. Adding to the problem are the relatively bright cluster stars: although none shine brighter than 8th magnitude, these stars turn into brilliant beacons when viewed through telescopes. The result is that glare can overpower your eye, rendering the nebula difficult to spot.

The nebula surrounding the cluster IC 1848 is another difficult test for backyard astronomers. Using the 13.1-inch reflector equipped with a 26mm Plossl eyepiece and a nebula filter, I observed a large, very faint, amorphous glow surrounding a fairly conspicuous star cluster. To my eye, the brightest section of nebulosity lies to the cluster's east. Like IC 1805, IC 1848 appears perfectly featureless and disappears from view without the filter in place. Look for it about 2.5° east of IC 1805 and about 4.5° north of Eta (η) Persei, the star at the tip of Perseus' head.

Now let's travel west to Gamma (γ) Cassiopeiae, the prototype for a class of rapidly rotating type-B variable stars. Marking the center of the constellation's W asterism, Gamma fluctuates erratically between magnitudes 1.6 and 3.0 and is associated with a pair of emission nebulae found 20' north and east of the star. These are designated **IC 59** and **IC 63**, respectively, and prove even more difficult to spy than either IC

NGC 1931
Size 3' by 3' by Martin C. Germano

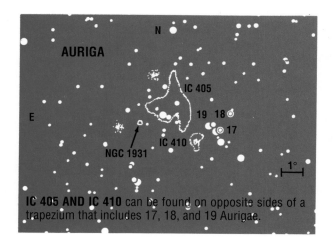

IC 405 AND IC 410 can be found on opposite sides of a trapezium that includes 17, 18, and 19 Aurigae.

1805 or IC 1848. In photographs both IC 59 and IC 63 appear as fan-shaped glows; visually, each is seen as little more than an indistinct glow.

IC 63 is the easier of the two to find. A 13.1-inch scope provides a hint of the object's cometlike shape, but the cloud's extremely low surface brightness makes its form difficult to confirm. IC 59 is tougher still. I think I saw it as a dim, shapeless glow surrounding a semicircular group of stars that reminded me of a miniaturized Corona Borealis. Once again, without the nebula filter in place, both objects vanished.

Perseus is home to **NGC 1579**. Begin your search at 2nd magnitude Zeta (ζ) Persei. Scanning 6° east-northeast, keep watch for three 6th-magnitude stars lined up in a tight north-south row. From this point, NGC 1579 is about 2° northwest. Through medium-sized amateur instruments, NGC 1579 appears as a small, faint cloud set just north of the 12th-magnitude star that illuminates it. NGC 1579 does not show a definite form through amateur scopes. Instead, most observers note it as irregular or perhaps rectangular in appearance. Photographs reveal its full 12' by 8' dimensions, as well as the presence of subtle dark lanes.

Three more challenging nebulae can be found in central Auriga, each within 3° of the 4th-magnitude star Phi (π) Aurigae. Brightest of the group is the emission nebula **NGC 1931,** located a little less than 1° east of Phi. Through an 8-inch telescope, NGC 1931 appears as a small, oblong mist of greenish light surrounding three faint stars. Larger instruments more readily define the nebula's unique shape — some observers have likened it to a peanut — as well as add a fourth star to the central trio.

Due west of Phi Aurigae is the strange variable star AE Aurigae. Studies show that AE, which usually maintains 6th magnitude, is an escapee from the Orion complex of nebulosity. The star has apparently been speeding on its way for about 2.7 million years. Other probable runaways from Orion are 53 Arietis and Mu Columbae. The proper motion of all three can be traced back to a common point near Eta (η) Orionis.

Surrounding AE Aurigae is a large emission nebula catalogued as **IC 405** and popularly known as the Flaming Star Nebula. In long-exposure photographs, IC 405 displays a striking filamentary structure that undoubtedly led to its incendiary nickname. Even through the largest amateur telescopes, IC 405 appears

IC 405 and IC 410
Sizes 30' by 19' and 40' by 30' by Chuck Vaughn

as a grayish puff of smoke set near a kite-shaped stellar asterism (AE Aurigae marks the top of the kite). The association of the variable star and IC 405 is a fleeting one, as the star is traveling through space about three times as fast as the nebula. As AE Aurigae passes through and then beyond the cloud, IC 405 will revert to a dark nebula.

The final complex of nebulosity in Auriga is the most difficult to spot. Engulfing the attractive open cluster NGC 1893, 2° southwest of Phi, is the large emission cloud **IC 410**. Although this object measures about 30' across, only a small portion of the nebula can be spotted visually. Look for a soft, irregular glow just northwest of the little cluster.

Bright nebulae offer great challenges to backyard astronomers, but when viewed through high-quality telescopes and under clear, dark skies, they are among the most rewarding. Don't have any reservations about searching for these challenging objects on a chilly winter night this year. Try your luck with these objects and they will not disappoint you. □

These objects are best visible in the winter evening sky.

Observing Bright Planetary Nebulae

All it takes is a small telescope to observe the brightest planetary nebulae — glowing shells of gas cast off by dying stars.

by Alan Goldstein

The universe is an awfully strange place. In it, matter is neither created nor destroyed but rather is continuously recycled by the deaths of old stars and births of new ones. A look at the lifestyles of stars — not of the rich and famous but of the hot and distant — reveals several ways in which the same material gets used over and over again.

Stars die in many ways, from cooling slowly like burning embers to violently exploding as supernovae. Somewhere in the middle lies the class of stars — including our Sun — that will calmly puff off shells of gas, forming what astronomers call planetary nebulae.

Observing these nebulae is a joy because many of them are bright and colorful, appearing as gray-green or blue-green disks. The challenge is not so much to find a planetary nebula, but to see the central star — the originator of the thin, luminescent gas cloud. The range of planetary nebulae runs from easy planetaries with faint central stars to faint nebular disks with bright central stars.

The prettiest planetary nebula in the sky also has a challenging central star. The **Ring Nebula** (M57, NGC 6720) in Lyra is easy to locate, lying between the bright stars Beta and Gamma Lyrae. In small telescopes, the Ring appears as a gray-green puff of smoke shaped like a ring. Larger scopes reveal a strong greenish color due to the presence of doubly ionized oxygen in the nebulosity. The perceived intensity varies from person to person and with differing telescope size. This nebula is a favorite of everybody. Beginners like it because it is easy to find and view, while most experienced observers never get tired of scrutinizing it to try to glimpse the central star.

M57's central star glows dimly at about magnitude 14.8. Theoretically, this places it within the range of a 12.5-inch telescope under ideal skies. Yet few observers have seen it in a telescope of that aperture. I have observed the Ring Nebula many times with a 21-inch telescope. On many nights the Ring's central star is invisible with this telescope. On the other hand, some keen-eyed observers claim to have seen the central star with a 5-inch telescope. Why such a discrepancy? There's no straight answer. It's due to a combination of factors, including experience, optics, transparency, the individual's sensitivity to blue light, and possibly the star itself. Because of the wide range of telescopes in which this star has been seen (and not seen), some astronomers believe it is a variable star.

NGC 6905 Mag. 11.9 Size 46" by Jack B. Marling

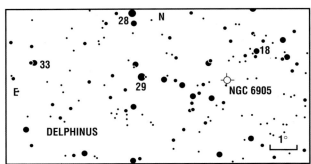

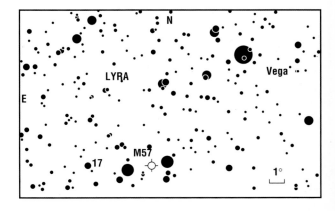

The Ring Nebula (M57)
Mag. 9.7   Size 70"   by Tony Hallas and Daphne Mount

The Dumbbell Nebula (M27)
Mag. 7.6   Size 350"   by Kim Zussman

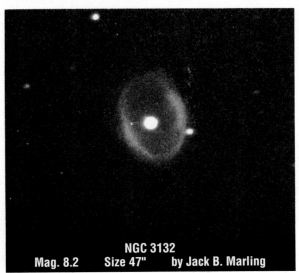

NGC 3132
Mag. 8.2   Size 47"   by Jack B. Marling

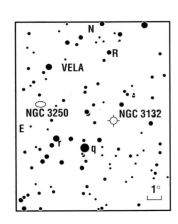

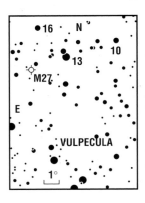

In addition to M57, we'll look at ten of the finest planetaries visible in the autumn sky. For convenience, we'll split them into three groups — one visible in the evening, another after midnight, and a third in the early morning sky.

The Evening Planetaries

The **Dumbbell Nebula** (M27, NGC 6853) in Vulpecula is a larger, brighter planetary than the Ring. The fact that M27 glows at magnitude 7.6 and measures 350" across puts it within easy reach of finderscopes and low-power binoculars. At about 50x, the oval disk appears to be evenly illuminated. With small telescopes, the notches in the dumbbell become easy. With larger instruments, the nebula takes on a mottled, patchy appearance. The distribution of the gas in the shell forming the Dumbbell Nebula is obviously not uniform in nature, but contains clumps of hot gas, hence the mottled appearance.

About half of the nebula has a sharp, distinct edge, or shock front. Superimposed on the nebular "disk" are numerous foreground stars. In the center lies the magnitude 13.9 bluish star that formed M27. This star is not too difficult in an 8-inch telescope under good skies and has been picked up with a 6-inch under superb atmospheric conditions.

Much smaller than either M57 or M27, **NGC 6210** is nonetheless the finest planetary nebula in Hercules. In most telescopes, NGC 6210 appears as a blue-green disk glowing at magnitude 9.3 with a central star about three magnitudes fainter. This planetary is almost a magnitude fainter than M57, yet its central star is much brighter. The planetary's size is tiny, however: with a diameter of 14", NGC 6210 would fill only the central hole in the Ring Nebula. Large backyard scopes reveal some detail around the perimeter of this nebula's disk. The central star is visible in all but the smallest telescopes. With small scopes it can be resolved from the glow of the nebula with high magnification.

NGC 6905 is a fairly bright planetary nebula in Delphinus that is slightly fainter than NGC 6210. NGC 6905 has a magnitude of 11.9 and a 14th-magnitude central star embedded in its mottled disk. A 10-inch scope may be necessary to pick out this faint blue star. If you can see the central star in M27 without difficulty, turn your telescope to NGC 6905 — averted vision may help, but it frequently brightens the nebula more than the central star.

Moving into the Night

As the stars wheel westward, and midnight has come and gone, it's time to try spotting **NGC 7662**, the brightest planetary nebula in Andromeda. It is a roundish object 20" in extent. Glowing green, NGC 7662 is a spectacular late-year object. It presents another great disparity. Glowing brightly at 9.2 magnitude, the nebula is easily visible in binoculars or small scopes, but the challenging central star glows feebly at magnitude 13.2. In theory, a good 8-inch or 10-inch telescope will show this star. However, it is especially difficult to see because it is involved with the nebulosity. I have never observed this central star. A filter that transmits blue light may be capable of dimming the nebular glow sufficiently to show the star. Good luck with this one!

NGC 246 Mag. 8.0 Size 225" by Jack B. Marling

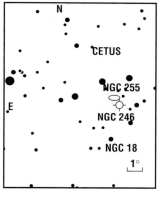

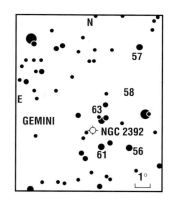

The Eskimo Nebula (NGC 2392) Mag. 9.9 Size 13" by Kim Zussman

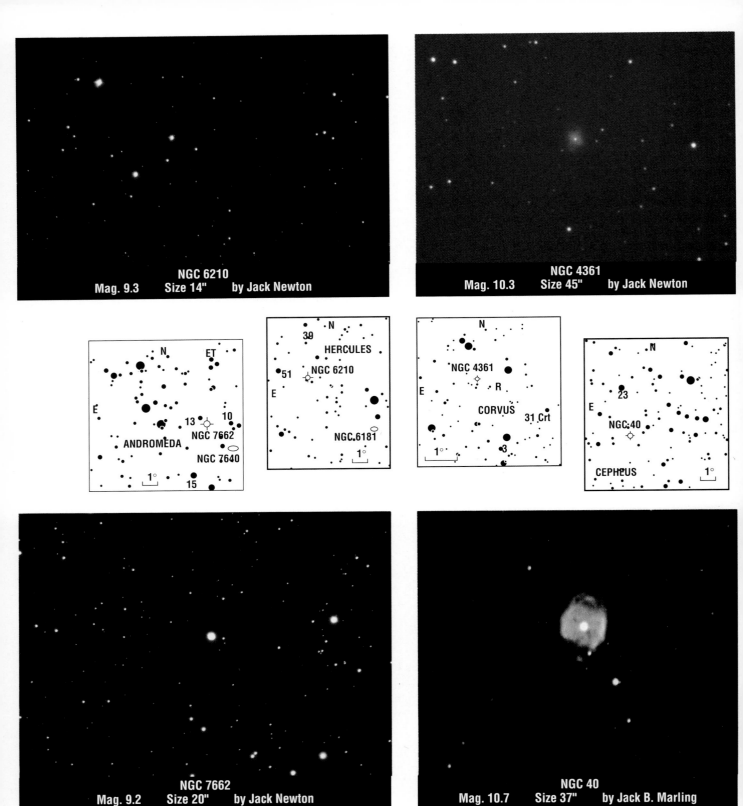

NGC 6210 — Mag. 9.3 — Size 14" — by Jack Newton

NGC 4361 — Mag. 10.3 — Size 45" — by Jack Newton

NGC 7662 — Mag. 9.2 — Size 20" — by Jack Newton

NGC 40 — Mag. 10.7 — Size 37" — by Jack B. Marling

NGC 40 is one of the brightest planetary nebulae in Cepheus. It is comparable to NGC 6905, glowing at magnitude 10.7. Its central star is similar to the one in NGC 6210, as it glows at magnitude 11.6. Both the nebula and the star are visible in small telescopes. The planetary is 37" across. Photographs show a somewhat squared-off ring with ansae (tenuously connected filaments or blobs of gas) along the long axis. The lower surface brightness of this nebula makes detail difficult to glimpse. The central star is not difficult with any telescope larger than 4 inches in aperture — a 3-inch telescope can detect the star under dark, transparent skies.

Visible in the southeastern sky after midnight, Cetus' **NGC 246** has a very low surface brightness. The central star is bright and conspicuous in small telescopes. The nebula's published magnitude of 8.0 is deceiving because this ghostlike object is definitely not

Bright Summer Planetaries

Object	R.A (2000.0)	Dec.	Mag.	Mag.*	Size	Con
NGC 40	0h13.0m	+72°32'	10.7	11.6	37"	Cep
NGC 246	0h47.0m	-11°53'	8.0	11.9	225"	Cet
NGC 2392	7h29.2m	+20°55'	9.9	10.5	13"	Gem
NGC 3132	10h07.7m	-40°26'	8.2	10.1	47"	Vel
NGC 3242	10h24.8m	-18°38'	8.6	12.0	16"	Hya
NGC 4361	12h24.5m	-18°48'	10.3	13.2	45"	Crv
NGC 6210	16h44.5m	+23°49'	9.3	12.9	14"	Her
M57	18h53.6m	+33°02'	9.7	14.8v?	70"	Lyr
M27	19h59.6m	+22°43'	7.6	13.9	350"	Vul
NGC 6905	20h22.4m	+20°07'	11.9	13.5B	46"	Del
NGC 7662	23h25.9m	+42°33'	9.2	13.2v?	20"	And

Object: Designation in Messier or NGC catalog. **R.A. and Dec.:** Coordinates in equinox 2000.0. **Mag.:** Photographic magnitude. **Mag.*:** V magnitude of central star; subscript B denotes blue magnitude. **Size:** Diameter in arcseconds. **Con.:** Constellation in which object lies.

The Ghost of Jupiter (NGC 3242)
Mag. 8.6   Size 16"   by Joe Liddell

easy in small telescopes. Even an 8-inch telescope requires good, dark skies to show NGC 246. The Ring Nebula, listed at nearly a full magnitude fainter, can be picked out under suburban sky glow, but M57 has a much higher surface brightness. At magnitude 11.9, the central star in NGC 246 is one of the brightest in the sky. A large instrument will reveal other stars superimposed on the somewhat annular gas shell. M27's disk is studded with stars, yet NGC 246 lies far from the Milky Way. Like the Dumbbell Nebula, this object contains irregular patches of bright gas. The unusual outline of the nebula may make the central star appear less "central" than it really is. What do you see in your telescope?

Before the Dawn

As the night continues, the bright constellations of winter will rise up over the eastern horizon. In this star-studded patch of sky you'll find another batch of bright planetaries. **NGC 2392** in Gemini is called the Eskimo Nebula because in long-exposure photos it resembles a face peering out from a fur-lined hood. Visually it is less graphic, appearing like a disk with a faint, outer rim. In large telescopes NGC 2392 looks like a double ring. The nebula itself has a visual magnitude of 9.9, the central star glows at magnitude 10.5. High magnification is necessary to distinguish the central star from the nebula. As is typical of high-contrast planetary nebulae, this object appears beautifully blue when viewed with large scopes.

NGC 3132 is a southern planetary in Vela just over the Vela/Antlia border. Nicknamed the Eight-Burst nebula, this object is a bright annular nebula similar to the Ring Nebula. At magnitude 8.2, NGC 3132 is easy to see in small telescopes. NGC 3132's southerly declination of −40° makes observing the planetary challenging for northern observers — look for it when it's near the meridian. At first glance, this object appears to have a bright central star. However, astronomers have found that the bright star appearing near the center of this object is not the progenitor star.

Located in Hydra is the bright, disklike planetary

NGC 3242. This fine object has been called the Ghost of Jupiter because its oval shape and 16" dimension remind observers of the solar system's largest planet. Most observers see NGC 3242 as greenish-blue or blue, more typical of Uranus or Neptune rather than Jupiter. The oval shape appears ringlike in large telescopes. The central star is a reasonably bright magnitude 12.0, putting it within reach of all but the smallest telescopes.

Our last planetary nebula is **NGC 4361** in Corvus. The diameter is a circular 45", somewhat smaller than the Ring Nebula. Very little detail is visible across the disk of NGC 4361 with either small or large telescopes. With instruments in the 4-inch to 6-inch range, it's best to look at NGC 4361 under very dark skies. The nebula's low contrast makes it extremely difficult in light-polluted skies. The central star should be detectable in a 4-inch scope under ideal conditions. An 8-inch instrument is necessary to see it with ease. The central star does not get lost in the glow of the nebula — as is the case with M57 and NGC 7662 — because of the very tenuous nature of the gas shell surrounding the central star.

This autumn try catching some planetary nebulae. Whether you're alone or at a star party, equipped with big telescopes or small, you can easily find planetaries and see nice detail within them. The opportunity to compare such objects as NGC 40 or NGC 3242 in a 3-inch, 8-inch and 16-inch telescope at the same time is extremely eye-opening. Whether you observe these planetaries with a small scope or a large one, you'll be sure to rate them high on your list of favorites in the autumn and winter skies.

These objects are best viewed at moderately high power on nights of steady seeing.

Observing Nebulosities in Cygnus

Challenge yourself to spot the North America Nebula, the Veil,
and the elusive clouds that envelop Gamma Cygni.
by Alan Goldstein

When you stand under the stars on a cool spring night, the rich starry realm of Cygnus the Swan lies overhead. The charm of Cygnus is simple: The Milky Way's shimmering path cuts directly through the constellation and, laced with dark voids and holes, connects it with Aquila to the south and Cepheus to the north. Cygnus, smack in the center of the northern Milky Way, is loaded with rich star fields, colorful double stars, open clusters, and nebulae of all shapes and sizes. But of all the riches of Cygnus the most alluring is its unique collection of nebulae. Nowhere else, not even in Sagittarius, do observers get the chance to see such a varied and unusual collection of glowing clouds of gas.

The largest and brightest nebula in Cygnus is the **North America Nebula** (NGC 7000). The object's curious nickname arose in the late nineteenth-century when German astronomer Max Wolf photographed this object and examined it carefully on a series of plates. NGC 7000 gets its shape from several large areas of dark nebulosity that surround it and obscure part of its face. This dark nebulosity creates several remarkable features similar to North American geography: The Gulf of Mexico, the Atlantic and Pacific oceans, Hudson Bay, and even Florida.

All of these features are distinctly visible on photographs made even with simple equipment, but observing them is somewhat more difficult. On dark, transparent nights the North America Nebula is visible to the naked eye and appears like an enormous bright glow southeast of Deneb, the brightest star in Cygnus. To the naked eye the shape of North America is somewhat recognizable, although not easily. However, under the same dark, transparent sky, low-power brings the nebula into sharp relief and shows the surrounding field crammed with bright stars.

A 3- or 4-inch telescope or large binoculars yields the best views of the North America Nebula because they provide modest magnification, a wide field of view, and good contrast. Larger rich-field instruments do not work as well as you'd think. While they show fainter stars, the field of view becomes narrower, the contrast of the nebula against the background Milky Way diminishes, and the great nebula becomes less impressive. The optimum field of view is about 3°, since the nebula is about one and a half degrees across. The reddish color visible in photos is not visible to the eye; visual observers see the nebula as gray-

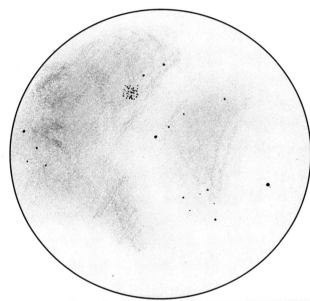

The North America Nebula
12.5-inch f/4.5 scope at 35x and 150x   by Jeff Corder

Central Cygnus
by Jack B. Marling

The North America Nebula
Size 120' by 100'   by Tony Hallas and Daphne Mount

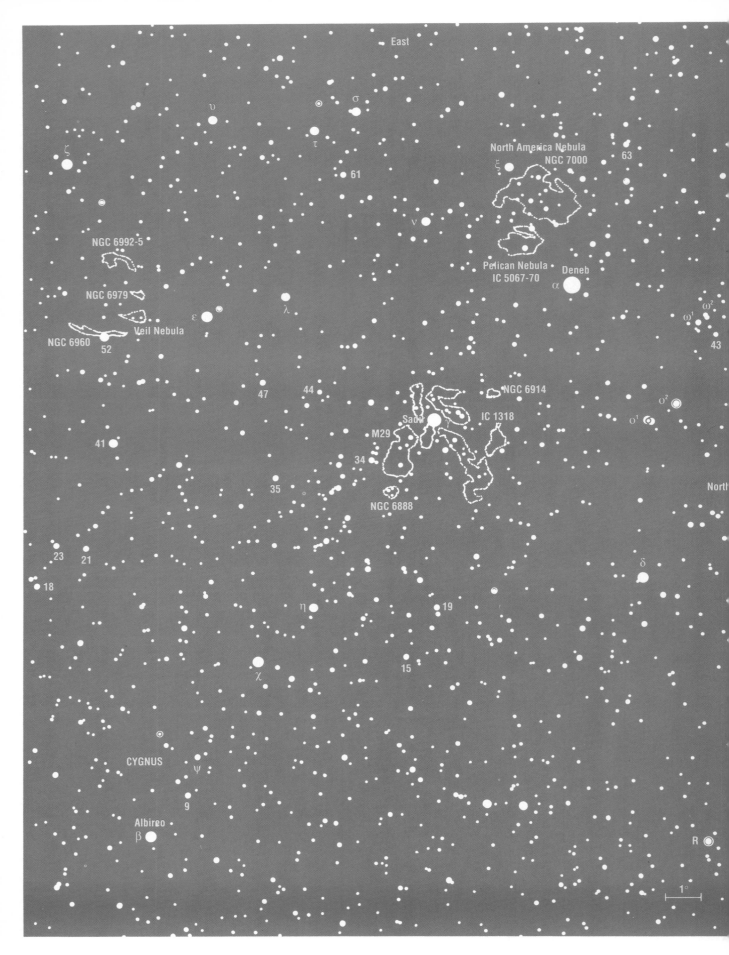

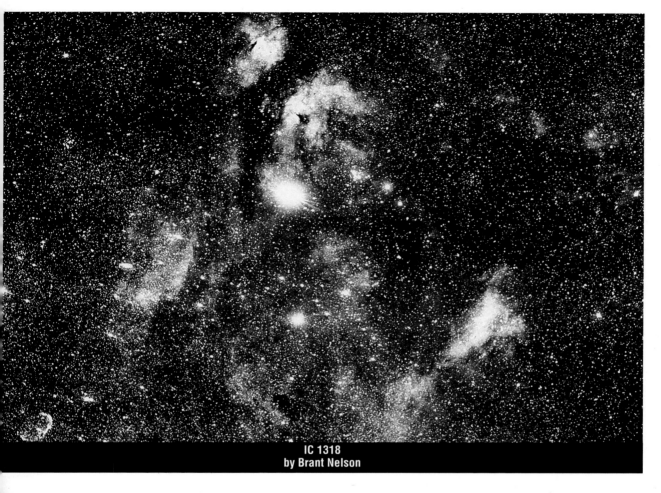

IC 1318
by Brant Nelson

The Pelican Nebula
Size 80' by 70'   by Walter E. Hamler

green. The most distinctive identifying features of the North America Nebula are the "Gulf of Mexico" and a dark, ring-shaped feature just south of "Hudson Bay." The stars inside the ring form a pattern that resembles a smiling face. Can you see this feature in your telescope?

Just west of the North America Nebula, about where Bermuda would be, lies the much fainter emission nebula called the **Pelican Nebula** (IC 5067-70). Its complex shape really does resemble a pelican, but only in photographs. Visually the object is difficult to see and appears as a large, oval glow with its brightest parts forming odd, curving shapes. Look for the gap in the pelican's lower bill and the bird's eye. Once identified from a photograph these features are unmistakable. The area representing the back of the pelican's head is the brightest part of the nebula. The same instrument requirements for the North America apply to the Pelican Nebula: low magnification, a wide field, and high contrast.

South of the Pelican Nebula lies IC 5068, a detached part of the nebular complex. It is as difficult to detect as the Pelican and appears as a series of wispy patches broken by dark nebulae and roughly aligned east-west.

Near the star Gamma (γ) in the center of Cygnus lies another complex bright-and-dark nebula. Designated **IC 1318**, this object consists of a group of large glowing clouds on three sides of Gamma Cygni separated by wide gulfs of obscuring material. The areas south and west of Gamma Cygni are particularly rich in nebulosity. IC 1318, like the Pelican Nebula, is

NGC 6888
Size 20' by 10'   by Scott Hicks

NGC 6914
Size 13' by 12'   by Martin C. Germano

NGC 6992
Size 60' by 8'   by Jack Newton

Nebulosities of Cygnus

Object	R.A. (2000.0)	Dec.	Mag.*	Size
NGC 6888	20h12.0m	+38°21'	7.4	20' by 10'
NGC 6960	20h45.7m	+30°43'	—	70' by 6'
NGC 6979	20h48.5m	+31°09'	—	45' by 30'
IC 5067-70	20h50.8m	+44°21'	—	80' by 70'
NGC 6992	20h56.4m	+31°43'	—	60' by 8'
NGC 6995	20h57.1m	+31°13'	—	12'
NGC 7000	20h58.8m	+44°20'	6.0	120' by 100'

Object: Designation in NGC or IC catalog. **R.A. and Dec.:** Right ascension and declination in equinox 2000.0. **Mag.*:** V magnitude of illuminating star. **Size:** Diameter in arcminutes.

much more difficult to see than the North America Nebula. Glancing off to the side of the field and employing your eye's light-sensitive rods, a technique called averted vision, will help you see faint detail in this nebulosity. As with the Pelican, short focal length telescopes in the 3- to 6-inch range work best. Larger scopes go fainter, but at the expense of seeing the complex in its entirety. Be particularly careful when you search for these low surface brightness objects. Several poor star clusters composed of faint stars lie in the area and, at first glance, masquerade as faint nebulosity.

A separate nebula lies about 2° north of Gamma Cygni. This object, **NGC 6914**, has a higher surface brightness and its 6' frame is rather easily visible in small telescopes on dark nights.

Near the southern tip of the IC 1318 complex lies **NGC 6888**, a peculiar filamentary nebula surrounding a Wolf-Rayet star. Whereas the North America and Pelican nebulae are stellar birthplaces in the process of creating clusters of stars, NGC 6888 is a cloud of glowing gas left from the dying puffs of its central star. NGC 6888 lies about a half degree southwest of RS Cygni and is recognizable by the diamond-shaped group of four stars within the nebula. This object is quite faint and measures 20' by 10', giving it a low surface brightness. However, under dark skies it is visible in small telescopes. At the Texas Star Party I have seen it with 7x50 binoculars. In small telescopes only the brightest parts of NGC 6888 are visible. In scopes larger than 12 inches, this object appears as an egg-shaped glow that is nearly a complete shell.

Similar to NGC 6888, Cygnus' grand **Veil Nebula** is one of the sky's greatest sights. Composed of several large strings of glowing light with the designations NGC 6960, NGC 6979, NGC 6992, and NGC 6995, this object is an ancient supernova remnant, the far-flung material from a star that exploded 50,000 years ago. The Veil Nebula covers an area nearly 3° across. Five Full Moons would fit comfortably within its diameter.

The Veil consists of three main elements: the brightest, eastern half, NGC 6992-5; the western half, NGC 6960; and the central section, NGC 6979. NGC 6992-5 is a thin, stringy glow oriented north-south that shows remarkable filamentary structure at high powers. NGC 6960 is a similar north-south nebula that passes behind the bright star 52 Cygni. NGC 6979 is a

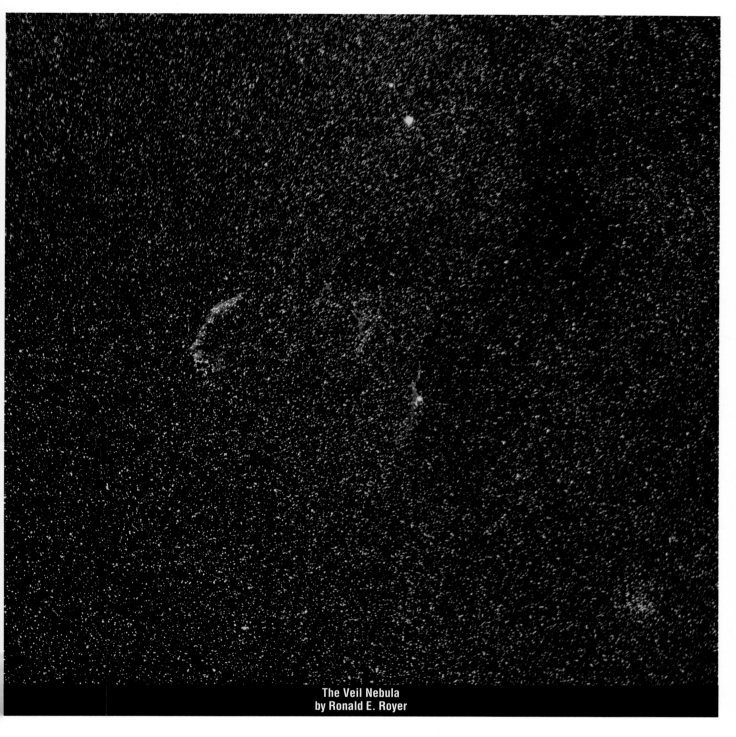

**The Veil Nebula
by Ronald E. Royer**

much fainter triangular nebulosity located in the northern central part of the object.

The best view of the entire object can be had with large binoculars on a very dark night. The two main components of the Veil appear as ghostly streaks set against an incredibly rich star field. To see details in the Veil, however, you must use a telescope. Small scopes show the object faintly but without much detail. Large telescopes, those 16 inches and up, show intricate details in the Veil. Under transparent skies, wispy, lacework structure is plainly visible in both main sections and is unforgettable in NGC 6992-5.

The famous and legendary emission nebulae in Cygnus will provide you with some of your most memorable deep-sky sights and some of your most extreme challenges. Study the positions of these objects carefully on star atlases and photographs, choose the right equipment and a steady, dark night. Above all, have patience. Before you know it, you'll be an old hand at exploring the Veil and charting the faintly recognizable coastline of a ghostly North America floating high overhead.

*These objects are best visible
in the summer evening sky.*

Great Summer Planetaries

The summer Milky Way is resplendent with bright planetary nebulae like
the Ring, the Dumbbell, and the Blinking Planetary.
by Alan Goldstein

Planetary nebulae have a lot to offer to backyard observers. These cosmic puffballs cover the entire spectrum of observational difficulty: Some are easily spotted with large binoculars, while others demand special filters and a big dose of patience to distinguish from the background stars. But regardless of whether you just want to marvel at the beauty of bright planetaries or you want the challenge of finding the faint ones, the best planetary nebulae can be found in the summer sky.

The finest planetary nebula in the sky is the **Ring Nebula** in Lyra (M57 or NGC 6720). The Ring is one of the most distinctively shaped of all deep-sky objects, although its appearance is something of an optical illusion. Not actually ring-shaped, it is a sphere of gas puffed away by its central star. It only looks like a ring because most of its light comes from the thick edges of the shell.

Although the Ring is bright, its small size (just over 1' in diameter) makes it difficult to distinguish from field stars when viewed with a 6x to 10x finder scope. In large binoculars or a 2-inch telescope at 20x to 50x, the Ring looks like a pale gray disk. A 3-inch telescope at 150x will show the Ring's distinctive shape, but only when you use averted vision. However, a 6-inch telescope's greater light grasp unmistakably shows the ring shape at 100x. In 8-inch and larger telescopes the Ring appears like an oval gray-green disk with a dark central hole. Still larger scopes reveal a dimly glowing haze within the central hole.

The most difficult observing challenge associated with the Ring Nebula is spotting its faint central star. Cataloged at magnitude 15.7, this white dwarf star is extremely elusive. Based solely on its magnitude, the central star should be visible in a 16-inch scope on a dark, transparent night. Some particularly keen-eyed observers with 5-inch telescopes have spotted the star. However, astronomers with 40-inch telescopes at professional observatories have sometimes missed seeing this star.

To explain this puzzle, astronomers have hypothesized that the Ring's central star varies in brightness. Also, its strong blue tint probably means that observers with eyes sensitive to blue light have a much easier time seeing the star than do observers whose eyes are more sensitive to red light. Nonetheless, the mystery of the Ring's central star will surely continue as one of the strangest stories in deep-sky observing.

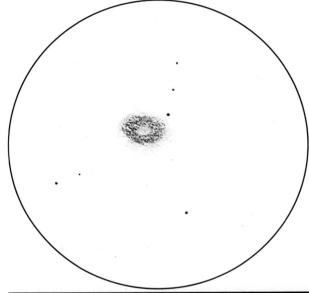

The Ring Nebula
8-inch f/10 SCT at 50x by David J. Eicher

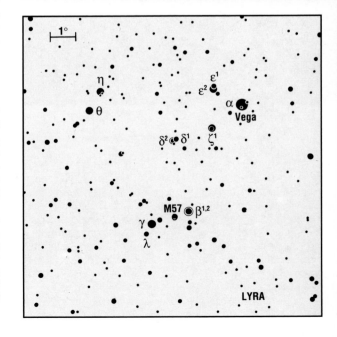

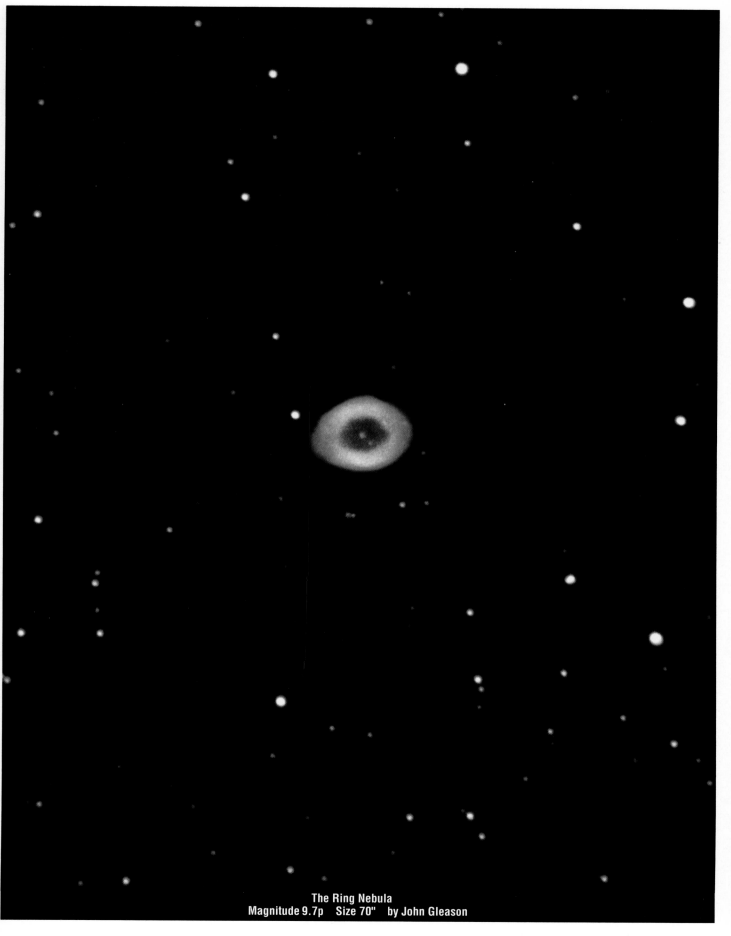

The Ring Nebula
Magnitude 9.7p   Size 70"   by John Gleason

The Dumbbell Nebula
Magnitude 7.6p   Size 350"   by Jack B. Marling

NGC 6905
Magnitude 11.9p   Size 46"   by Jack B. Marling

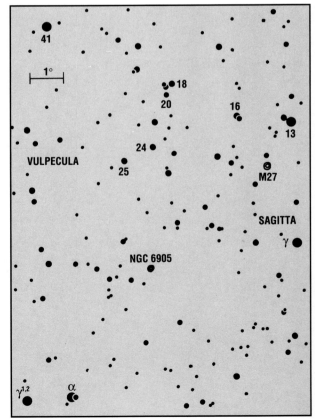

NGC 7008
Magnitude 13.3p Size 83" by Jack B. Marling

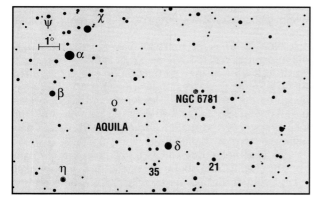

Second only to the Ring in visual appeal is the **Dumbbell Nebula** in Vulpecula (M27 or NGC 6853). The Dumbbell can be found by scanning 2° due north of Gamma Sagittae to a C-shaped group of stars about 30' south of 14 Vulpeculae. The Dumbbell Nebula lies in the center of this distinctive star group.

Named for its unique shape, the Dumbbell is visible through a 3-inch telescope at 150x as a disk with two giant bites out of it. (Some observers think it resembles a butterfly more than a dumbbell.) Like the Ring Nebula, the Dumbbell has a high surface brightness and holds up well under high magnification. If the night is steady and dark, don't be afraid to use 200x or more to see detail in the Dumbbell.

On first viewing this object you'll see that it is not uniformly illuminated across its disk, which gives it a patchy appearance. Several foreground stars are visible superimposed on the face. The Dumbbell's central star glows at magnitude 13.5, making it visible with an 8-inch scope.

Northward into Cygnus, **NGC 6826** can be found less than 5' east of the 6th-magnitude star 16 Cygni. With a magnitude of 9.8 and a central star shining at magnitude 9.9, NGC 6826 is difficult to miss in a small telescope. It is rather small, measuring 30" across, so be sure to use at least 100x when you zero in on the nebula's position. When you come across the right field you'll see that NGC 6826 has a bright, nonstellar blue-green disk. You may see the central star immediately when you scan across the nebula.

The visual relationship between nebula and central star is most unusual in the case of NGC 6826. When you look directly at NGC 6826, the nebula disappears and you see only the central star. When you look slightly to one side of the field, however, the star disappears and the nebula pops into view. This peculiar effect gives NGC 6826 its nickname, the Blinking Planetary.

Another wonderful Cygnus planetary is **NGC 7008**, a large, ghostly shell of light. Located in the northernmost reaches of this constellation close to the Cepheus border, NGC 7008 lies midway between Mu Cephei and Omega^2 Cygni. Its diameter of 83" makes it one of the largest planetary nebulae in the summer sky.

Most sources assign NGC 7008 a magnitude of 13, so in a small telescope this object appears large and diffuse with a low surface brightness. The best views of this object come with a 10-inch or larger scope at 100x. The nebula's central star, a 12th-magnitude dot, is within the reach of 6-inch telescopes under dark skies. NGC 7008 has an irregular

NGC 6781
Magnitude 11.8p Size 109" by Jack Newton

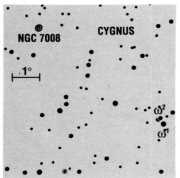

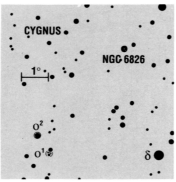

NGC 6826
Magnitude 9.8p Size 30" by Lee C. Coombs

Great Summer Planetaries

Object	R.A. (2000.0) Dec.		Mag.	Size
M57	18h53.6m	+33°02'	9.7p	70"
NGC 6781	19h18.4m	+06°33'	11.8p	109"
NGC 6826	19h44.9m	+50°31'	9.8p	30"
M27	19h59.6m	+22°43'	7.6p	350"
NGC 6905	20h22.4m	+21°06'	11.9p	46"
NGC 7008	21h00.6m	+54°33'	13.3p	83"

Object: Designation in Messier or NGC catalog. **R.A. and Dec.:** Right ascension and declination in equinox 2000.0. **Mag.:** Photographic magnitude. **Size:** Diameter in arcseconds.

but basically oval shape that is visible in telescopes as small as 8 inches in aperture.

NGC 6905 is a bright planetary nebula in the small constellation Delphinus. Lying 2° southwest of the 6th-magnitude star 18 Sagittae, this nebula is large enough (46") and bright enough (12th magnitude) to be picked up in a 3-inch telescope with little trouble. Its oval shape becomes apparent in 6-inch scopes at 100x. This object bears magnification well, so use a relatively high-power eyepiece for the best view.

NGC 6905's central star glows dimly at about 14th magnitude and is invisible in telescopes smaller than 10 inches in aperture. I've always seen NGC 6905 as gray in color, although some observers have recorded a greenish or even bluish hue, especially when using telescopes larger than 16 inches in aperture.

NGC 6781 is the most spectacular planetary nebula in Aquila. You can locate NGC 6781 by moving 2° north-northeast of the 5th-magnitude star 22 Aquilae. At magnitude 11.8, this large, ring-shaped object can be glimpsed in a 4-inch telescope. NGC 6781 measures 109" across, which gives it a relatively low surface brightness and makes finding it on less than dark nights somewhat challenging.

In small telescopes NGC 6781 appears as a pale, ghostly disk. Its ring shape becomes more pronounced as you view it through larger telescopes. Some observers have seen a ringlike shape with 8-inch telescopes. In larger apertures, the ring-shape is easy.

Like M57, Aquila's ring nebula is not symmetrical. The central "hole" is offset from the center of the circular disk, giving NGC 6781 the appearance of a ring that is thicker on one side than the other. The central star dimly glows at magnitude 15.4 and is visible only with the largest amateur telescopes.

Bright and faint, large and small, planetary nebulae offer a rich variety of shapes and details to challenge your observing skills. Why not unlimber your telescope and head out under the stars this summer when the best and the most challenging nebulae lie waiting overhead every night? □

These objects are best visible in the summer evening sky.

Bibliography

The following 10 titles are heartily recommended for beginning deep-sky observers. The bibliography that follows lists 60 other works useful for surveying clusters, galaxies, and nebulae with telescopes.

References for Beginners

ASTRONOMY, Kalmbach Publishing Co., Waukesha, Wisconsin. Founded in 1973, this monthly is the largest English-language astronomy periodical. It regularly contains plentiful information about deep-sky objects.

Berry, Richard. *Discover the Stars.* 119 pp., paper. Harmony Books, New York, 1987. The former editor-in-chief of ASTRONOMY introduces naked-eye, binocular, and small telescope observing using twelve all-sky maps and twenty-three close-up maps.

Consolmagno, Guy, and Dan M. Davis. *Turn Left at Orion; a hundred night sky objects to see in a small telescope — and how to find them.* 205 pp., hardcover. Cambridge University Press, New York, 1990. A brief description, sketch, and finder chart for one hundred objects of unusual interest to amateur observers.

Dickinson, Terence, and Alan Dyer. *The Backyard Astronomer's Guide.* 295 pp., hardcover. Camden House Publishing Co., Camden East, Ontario, 1991. A luxuriously illustrated beginner's guide to the hobby of amateur astronomy.

Eicher, David J. *Beyond the Solar System; 100 best deep-sky objects for amateur astronomers.* 80 pp., paper. AstroMedia, a division of Kalmbach Publishing Co., Waukesha, Wisconsin, 1992. An introduction to the brightest and most spectacular of the sky's nebulae, clusters, and galaxies.

Jones, Kenneth Glyn. *Messier's Nebulae and Star Clusters.* Second ed., 427 pp., hardcover. Cambridge University Press, New York, 1991. One of England's foremost amateur astronomers presents descriptions and eyepiece drawings for each of the Messier objects.

Karkoschka, Erich. *The Observer's Sky Atlas.* 130 pp., paper. Springer-Verlag, New York, 1990. A wonderful pocket-sized star atlas showing enough detail to find bright deep-sky objects.

Newton, Jack, and Philip Teece. *The Guide to Amateur Astronomy.* 327 pp., hardcover. Cambridge University Press, New York, 1988. The best all-around introduction to what amateur astronomy is all about.

Tirion, Wil. *Sky Atlas 2000.0.* Twenty-six fold-out folio charts, spiral-bound. Cambridge University Press and Sky Publishing Corp., New York, 1981. A large-scale atlas showing 43,000 stars down to magnitude 8 and 2,500 deep-sky objects in color.

Vehrenberg, Hans. *Atlas of Deep-Sky Splendors.* Fourth ed., 246 pp., hardcover. Treugesell-Verlag and Sky Publishing Corp., Dusseldorf, 1981. A splendid photographic album containing images of hundreds of deep-sky objects all reproduced at the same scale for easy comparison.

Other Works

Arp, Halton C. *Quasars, Redshifts, and Controversies.* 198 pp., hardcover. Interstellar Media, Berkeley, California, 1987. Semitechnical account of how a respected astronomer has found certain galaxies with anomalous redshifts.

Arp, Halton C., and Barry F. Madore. *A Catalogue of Southern Peculiar Galaxies and Associations.* Two vols., 208 pp., and 1,017 black-on-white photographs. Cambridge University Press, New York, 1987. A marvelous atlas of peculiar galaxies in the southern sky, giving the reader a great appreciation of the variety of galaxy forms.

Audouze, Jean, and Guy Israel, eds. *The Cambridge Atlas of Astronomy.* Second ed. 432 pp., hardcover. Cambridge University Press, New York, 1988. A folio-sized, colorful survey of astronomy that contains much fundamental information useful for backyard astronomers.

Barnard, Edward Emerson. *Atlas of Selected Regions of the Milky Way.* Two vols., 51 maps and 51 plates, hardcover. Carnegie Institution of Washington, Washington, D.C., 1927. The landmark atlas showing dark nebulae distributed throughout the Milky Way. Consists of original photographic prints pasted into an atlas.

Berendzen, Richard, Richard Hart, and Daniel Seeley. *Man Discovers the Galaxies.* 228 pp., paper. Columbia University Press, New York, 1984. Three distinguished historians present the tale of how early scientists unraveled the mystery of the Milky Way Galaxy.

Bok, Bart J., and Priscilla F. Bok. *The Milky Way.* Fifth ed., 356 pp., hardcover. Harvard University Press, Cambridge, Massachusetts, 1981. *The Milky Way* is generally regarded as the definitive book on the subject for nontechnical audiences.

Burnham, Robert. *The Star Book.* 17 pp., spiral-bound. AstroMedia and Cambridge University Press, Milwaukee, 1983. Group of seasonal star maps from ASTRONOMY magazine with a brief discussion of what to see.

Burnham, Robert, Jr. *Burnham's Celestial Handbook.* Three vols., 2,138 pp., paper. Dover Publications, New York, 1978. This voluminous compilation of deep-sky objects contains many photographs, charts, and tables.

Clark, Roger Nelson. *Visual Astronomy of the Deep Sky.* 355 pp., hardcover. Cambridge University Press, New York, and Sky Publishing Corp., Cambridge, Massachusetts, 1991. Photographs and sketches side-by-side illustrate an array of deep-sky objects in this guidebook.

Corwin, Harold R., Jr., Antoinette de Vaucouleurs, and Gerard de Vaucouleurs. *Southern Galaxy Catalogue.* 308 pp., paper. The University of Texas Press, Austin, 1985. This valuable catalog contains fundamental data for 5,841 galaxies south of declination -17°, assembled from studies of photographs made with the U.K. Schmidt telescope.

Couteau, Paul. *Observing Visual Double Stars.* 257 pp., paper. The Massachusetts Institute of Technology Press, Cambridge, Massachusetts, 1982. A studious French observer's excellent introduction to following double stars.

Covington, Michael. *Astrophotography for the Amateur.* 168 pp., hardcover. Cambridge University Press, New York, 1985. Chapters provide how-to advice for getting started in astrophotography.

de Vaucouleurs, Gerard, Antoinette de Vaucouleurs, Harold R. Corwin, Jr., Ronald J. Buta, Georges Paturel, and Pascal Fouque. *Third Reference Catalogue of Bright Galaxies.* Three vols., 2,069 pp., hardcover. Springer-Verlag, New York, 1991. The most valuable modern compilation of data for galaxies, covering 23,024 objects.

Dixon, Robert S., and George Sonneborn, compilers. *A Master List of Nonstellar Optical Astronomical Objects.* 835 pp., hardcover. Ohio State University Press, Columbus, 1980. An all-in-one compilation of more than 185,000 deep-sky objects drawn from 270 catalogues. This list is invaluable for identifying objects.

Dreyer, John Louis Emil. *New General Catalogue of Nebulae and Clusters of Stars (1888). Index Catalogue (1895). Second Index Catalogue (1908).* 378 pp., paper. Royal Astronomical Society, London, 1962. This single volume presents a facsimile reprint of the original basic listings of deep-sky objects compiled by Dreyer.

Eicher, David J., and the editors of *Deep Sky* magazine. *Deep Sky Observing with Small Telescopes.* 331 pp., paper. Enslow Publishers, Hillside, New Jersey, 1989. A beginner's manual for observing deep-sky objects with 2-inch to 6-inch telescopes. Contains extensive listings of objects and many photographs and eyepiece sketches made by backyard observers.

Eicher, David J. *The Universe from Your Backyard.* 188 pp., hardcover. Cambridge University Press and AstroMedia, a division of Kalmbach Publishing Co., New York, 1988. This book is a series of republished "Backyard Astronomer" articles from ASTRONOMY magazine. Included in its coverage are forty-six constellations or groups of constellations and 690 deep-sky objects. A three-color map, eyepiece sketches, and color photographs appear for each constellation.

Ferris, Timothy. *Galaxies.* 191 pp., hardcover. Stewart, Tabori & Chang, New York, 1980. A folio-sized photo essay on galaxies with an engagingly written narrative.

Hartung, E.J. *Astronomical Objects for Southern Telescopes.* 238 pp., hardcover. Cambridge University Press, New York, 1968. Meticulous observing notes by an Australian observer for many deep-sky objects in the Southern Hemisphere.

Hirshfeld, Alan, and Roger W. Sinnott, eds. *Sky Catalogue 2000.0.* Cambridge University Press and Sky Publishing Corp., New York, 1982-1985. Two vols. Vol. 2 (356 pp., hardcover) lists fundamental data for thousands of double and variable stars, 750 open clusters, 150 globular clusters, 238 bright nebulae, 150 dark nebulae, 564 planetary nebulae, 3,116 galaxies, and 297 quasars.

Hodge, Paul. *Atlas of the Andromeda Galaxy.* 79 pp., hardcover. The University of Washington Press, Seattle, 1982. A single-volume set of photographic maps that display all of the individual features in M31.

Hodge, Paul. *Galaxies.* 174 pp., hardcover. Harvard University Press, Cambridge, Massachusetts, 1986. A revision of the classic introduction to galaxy research written by Harlow Shapley.

Jones, Kenneth Glyn. *The Search for the Nebulae.* 84 pp., hardcover. Alpha Academic, Giles, England, 1975. A compendium of the recorded notes about deep-sky objects by several dozen early observers.

Jones, Kenneth Glyn, ed. *The Webb Society Deep Sky Observer's Handbook.* Eight vols., 1,544 pp., paper. Enslow Publishers, Hillside, New Jersey, 1979-1990. Collection of observations of deep-sky objects by amateur astronomers, made with everything from binoculars to large, professional telescopes. Covers double stars (vol. 1), planetary and gaseous nebulae (vol. 2), open and globular clusters (vol. 3), galaxies (vol. 4), clusters of galaxies (vol. 5), anonymous galaxies (vol. 6), Southern Hemisphere objects (vol. 7), and variable stars (vol. 8).

Lauberts, Andris. *The ESO/Uppsala Survey of the ESO (B) Atlas.* 503 pp., hardcover. European Southern Observatory, Munich, 1982. A massive data compilation containing fundamental information for 18,438 objects in the Southern Hemisphere. The listing results from an effort to photograph the southern sky in blue light.

Laustsen, Svend, Claus Madsen, and Richard M. West. *Exploring the Southern Sky; a pictorial atlas from the European Southern Observatory (ESO).* 274 pp., hardcover. Springer-Verlag, New York, 1987. This work consists of a long string of captions about a stunning collection of some of the world's finest color astrophotography.

Levy, David H. *Observing Variable Stars; a guide for the beginner.* 180 pp., hardcover. Cambridge University Press, New York, 1989. A thorough introduction and observing guide containing data on all major variable stars and other variable objects such as comets, active galaxies, and the Sun.

Levy, David H. *The Sky; a user's guide.* 295 pp., hardcover. Cambridge University Press, New York, 1991. An introduction to the various phenomena visible in the night sky.

Liller, Bill, and Ben Mayer. *The Cambridge Astronomy Guide.* 176 pp., hardcover. Cambridge University Press, New York, 1985. Alternating chapters by professional and amateur astronomers give the reader a feeling of what it's like to begin observing the sky.

Luginbuhl, Christian B., and Brian A. Skiff. *Observing Handbook and Catalogue of Deep-Sky Objects.* 352 pp., hardcover. Cambridge University Press, New York, 1990. The authoritative single volume for data on deep-sky objects, this book contains information on nearly 2,050 galaxies, nebulae, and clusters.

Malin, David, and Paul Murdin. *Colours of the Stars.* 198 pp., hardcover. Cambridge University Press, New York, 1984. Professional astronomers demonstrate the hows and whys of color in astronomy and what it reveals about the universe. Contains numerous color photographs of deep-sky objects.

Mallas, John H., and Evered Kreimer. *The Messier Album.* 216 pp., hardcover. Sky Publishing Corp., Cambridge, Massachusetts, 1978. A photograph, sketch, and brief description for each Messier object.

Mayer, Ben. *Astrowatch.* 140 pp., paper. Perigee Books, New York, 1988. Find the constellations by using a metal frame and series of templates.

Menzel, Donald, and Jay M. Pasachoff. *A Field Guide to the Stars and Planets.* 473 pp., paper. Houghton-Mifflin, Boston, 1983. Pocket-sized introduction to observing the sky with binoculars.

Mitton, Simon. *Exploring the Galaxies.* 206 pp., paper. Charles Scribner's Sons, New York, 1976. A semitechnical discussion of galaxies and galaxy research.

Mitton, Simon, ed. *The Cambridge Encyclopaedia of Astronomy.* 456 pp., hardcover. Crown Publishers, New York, 1977. Authoritative, fact-filled reference tome covering all aspects of modern astronomy.

Murdin, Paul, and David Allen. *Catalogue of the Universe.* 256 pp., hardcover. Crown Publishers, New York, 1979. Description of bright and unusual objects inhabiting the universe, sprinkled with photographs throughout.

Neckel, Thorsten, and Hans Vehrenberg. *Atlas of Galactic Nebulae.* Two vols., 286 pp., spiral-bound. Treugesell-Verlag, Dusseldorf, West Germany, 1985-1987. An extensive atlas of negative plates of nebulae collected from a variety of sources, printed on heavy card stock.

Newton, Jack. *Deep Sky Objects; a guide for the amateur astronomer.* 160 pp., hardcover. Gall Publications, Toronto, 1977. A Canadian astrophotographer's compilation of photographs of each of the Messier objects and a few NGCs.

Newton, Jack, and Philip Teece. *The Cambridge Deep-Sky Album.* 126 pp., hardcover. Cambridge University Press and AstroMedia Corp., New York, 1984. This work contains color photographs of all the Messier objects and many NGC objects.

Parker, Sybil P., ed. *The McGraw-Hill Encyclopedia of Astronomy.* 450 pp., hardcover. McGraw-Hill, New York, 1983. An extremely useful reference, with helpful entries about all aspects of astronomy.

Peltier, Leslie C. *Leslie Peltier's Guide to the Stars.* 185 pp., paper. AstroMedia Corp. and Cambridge University Press, Waukesha, Wisconsin, 1986. A basic introduction to observing stars, planets, and deep-sky objects with binoculars.

Peltier, Leslie C. *Starlight Nights.* 236 pp., paper. Sky Publishing Corp., Cambridge, Massachusetts, 1965. One of the greatest amateur astronomers of all time describes his experiences as a stargazer.

Ridpath, Ian. *The Night Sky.* 240 pp., paper. Collins Gem Guides, London, 1985. A micro-sized (3 1/4" by 4 5/8") set of star maps by Wil Tiron and accompanying list of objects. It's a perfect pocket accessory for observing.

Ridpath, Ian, ed. *Norton's 2000.0. Star Atlas and Reference Handbook.* Eighteenth ed., 179 pp. + 16 charts, hardcover. Longman Scientific and Technical and John Wiley and Sons, New York, 1989. A classic, *Norton's* provides charts showing stars down to magnitude 6 and an introductory discussion about observing.

Sandage, Allan. *The Hubble Atlas of Galaxies.* Fifty plates and accompanying text, hardcover. Carnegie Institution of Washington, Washington, D.C., 1961. Magnificent atlas of galaxies containing black-on-white photographs made with the world's finest telescopes.

Sandage, Allan, and John Bedke. *Atlas of Galaxies Useful for Measuring the Cosmological Distance Scale.* 13 pp. + 95 photographic panels, hardcover. National Aeronautics and Space Administration, Washington, 1988. This magnificent volume contains poster-sized black-on-white prints of nearby galaxies.

Sandage, Allan, and Gustav A. Tammann. *A Revised Shapley-Ames Catalog of Bright Galaxies.* 157 pp., hardcover. Carnegie Institution of Washington, Washington, D.C., 1987. Update of a famous study of 1,246 bright galaxies completed by the Harvard astronomers Shapley and Ames in 1932. Includes an atlas of black-on-white photographs of eighty-four galaxies in the survey.

Sanford, John. *Observing the Constellations; an A-Z guide for the amateur astronomer — fully illustrated with star maps and photographs.* 176 pp., paper. Simon & Schuster, New York, 1989. A constellation-by-constellation guide containing brief descriptions, photographs, and maps of the most inviting objects for backyard telescopes.

Sinnott, Roger, ed. *NGC 2000.0; the complete New General Catalogue and Index Catalogues of Nebulae and Star Clusters by J.L.E. Dreyer.* 273 pp., paper. Sky Publishing Corp. and Cambridge University Press, Cambridge, Massachusetts, 1988. A computer-generated revision of the *NGC* and *IC* catalogs with revised data for each object listed and coordinates given for equinox 2000.0. A valuable addition to the deep-sky observer's bookshelf.

Sky & Telescope. Sky Publishing Corp., Cambridge, Massachusetts. The oldest astronomy magazine in America, *Sky & Telescope* contains a monthly "Deep Sky Wonders" column written by the experienced observer Walter Scott Houston.

Sulentic, Jack W., and William G. Tifft. *The Revised New General Catalogue of Nonstellar Astronomical Objects.* 383 pp., hardcover. The University of Arizona Press, Tucson, 1973. The *Revised New General Catalogue* is an update of the standard list of bright deep-sky objects compiled by William and John Herschel. It contains positions, magnitude, and brief encoded descriptions for over 7,800 objects.

Thompson, Gregg D., and James T. Bryan, Jr. *The Supernova Search Charts and Handbook.* 134pp., hardcover, + 100 large quarto charts, boxed. Cambridge University Press, New York, 1990. An important set that constitutes the finest easily available resource for supernova hunting.

Tirion, Wil, Barry Rappaport, and George Lovi. *Uranometria 2000.0.* Two vols., 473 quarto-sized charts, hardcover. Wilmann-Bell, Inc., Richmond, Virginia, 1987-1988. A minutely detailed, large-scale atlas, *Uranometria 2000.0* shows 332,556 stars down to magnitude 9.5 and many thousands of deep-sky objects.

Tully, R. Brent. *Nearby Galaxies Catalogue.* 214 pp., hardcover. Cambridge University Press, New York, 1987. Contains fundamental data for 2,367 galaxies plotted in the companion work, the *Nearby Galaxies Atlas.*

Tully, R. Brent, and J. Richard Fisher. *Nearby Galaxies Atlas.* Twenty-two plates, spiral bound. Cambridge University Press, New York, 1987. A visually stunning atlas showing color reproduction of two- and three-dimensional maps of the galaxies closest to home.

Vehrenberg, Hans, and Dieter Blank. *Handbook of the Constellations.* Fifth ed., 197 pp., hardcover. Treugesell-Verlag, Dusseldorf, 1987. A constellation-by-constellation listing of bright stars and deep-sky objects, neatly fitted about a map showing each star group.

Wallis, Brad D., and Robert Provin. *A Manual of Advanced Celestial Photography.* 388 pp., hardcover. Cambridge University Press, New York, 1988. Describes sophisticated techniques for accomplished sky photographers.

Webb, Thomas W. *Celestial Objects for Common Telescopes.* Vol. 2, 351 pp., paper. Dover Publications, New York, 1962. A reprinting of an 1859 work that contains historically interesting accounts of double stars and nebulae.

Wray, James D. *The Color Atlas of Galaxies.* 189 pp., hardcover. Cambridge University Press, New York, 1988. A spectacular tour of galactic forms, this McDonald Observatory astronomer's work displays over six hundred galaxies in color.

The Authors

Alan Goldstein became interested in astronomy in 1973 and joined the Louisville Astronomical Society one year later. In 1976 he was a cofounder of the National Deep Sky Observers Society, an observing group for which he serves as National Coordinator. He served as a contributing editor of *Deep Sky* magazine. By profession Goldstein is Earth and Natural Sciences Coordinator and Collections Curator at the Museum of History and Science in Louisville. In addition to writing about astronomy, he frequently writes about mineralogy, paleontology, and the history of science.

A native New Yorker, **Phil Harrington** became interested in astronomy in his 6th-grade science class in 1968. He became actively involved with observing using binoculars and an 8-inch telescope. In 1970 he began an annual trek to the Stellafane convention. Harrington has worked at the American Museum-Hayden Planetarium in New York City and is currently a mechanical engineer, although he teaches classes in observational astronomy at the Vanderbilt Planetarium in Centerport, New York. He is the author of many articles in ASTRONOMY, *Deep Sky*, and *Sky & Telescope*, and in 1990 his first book was published, *Touring the Universe with Binoculars* (John Wiley and Sons, New York).

A 20-year veteran of the military, **David Higgins** is a Senior Advisor for a U.S. Army Reserve Unit in Texas, overseeing the successful completion of duties assigned to infantry battalions. Born in Oregon, Higgins became interested in astronomy as a child and especially after his father gave him a 2-inch refractor. He now uses a 24-inch reflector to observe a wide array of deep-sky objects, concentrating on globular star clusters and dark nebulae. He was a contributing editor of *Deep Sky* magazine, and has written extensively for ASTRONOMY. "I've been observing for more than 25 years from Saigon to Vietnam jungles to the deserts of Ethiopia to western Texas," says Higgins. "I believe I enjoy Texas best, since one has to worry only about stars — not artillery fire."

Richard W. Jakiel is a native New Yorker who is a PhD student in geochemistry at Georgia Tech University. He became interested in astronomy in the early 1970s and, after buying an 8-inch reflector, was "hooked" on deep-sky observing. He has sketched galaxies and other deep-sky objects extensively and has a wide range of experience with telescopes. "I became involved with science education at Fredonia State University in New York and Northern Illinois University, where I went to college and then to graduate school," says Jakiel. "My experience with the Buffalo Astronomical Society and the Atlanta Astronomical Society allows me to stay in contact with other amateurs and do frequent deep-sky observing."

A meteorologist by profession, **Alister Ling** delights in many aspects of the sky, night and day. He has a great interest in the unusual — whether it's detecting filaments of nearly invisible nebulae with an O III nebular filter and homemade 12.5-inch telescope, or observing fascinating detail within Messier objects. When clouds interfere with astronomy, his curiosity is heightened by anything from wispy "mare's tails" to the incredible structure and power of a severe thunderstorm. Alister willingly shares his 15 years of observing experience through talks, newsletters, and magazine articles, including his "Small Scope Showcase" column in *Deep Sky,* of which he served as a contributing editor.

Chicago's **Steve Lucas** is a truck driver for a freight company who, in 1985, founded the sole network of amateur astronomers devoted to searching for supernovae in other galaxies. Called SUNSEARCH, Lucas' program ties together 200 observers from around the globe who constantly monitor "program objects" for newly visible stars. He is editor and publisher of *Starburst*, a newsletter that informs readers about the latest supernova discoveries and recent observations of supernovae in other galaxies.

Rod Pommier is a surgical oncologist and assistant professor of surgery at the Oregon Health Sciences University in Portland. He has been an amateur astronomer since age 12 and is an avid astrophotographer. Pommier is a member of the Amateur Astronomers Association of New York, the Rose City Astronomers, and the Friends of Pine Mountain Observatory. He has published articles about astrophotographic techniques, and some of his astrophotos hang in the Cornell University Medical Library in New York City.

Max Radloff is a classical pianist who lives with his wife Leslie in South St. Paul, Minnesota. His astronomical interests include deep-sky and planetary observing, telescope making, and helping to run the Minnesota Astronomical Society, of which he is secretary. He has written articles for ASTRONOMY and *Deep Sky*. In addition to astronomy Radloff enjoys concerts, cross-country skiing, and cooking oriental food.

A research and development engineer for an electronics firm in Scottsdale, Arizona, **Chris Schur** is also one of the most respected astrophotographers in the world. Originally from Michigan, Schur became intensely interested in astronomy after moving to Arizona in 1979. He frequently writes about astrophotography. "I've found during recent years that I greatly enjoy shooting emission nebulae and comets," says Schur. "I use both a 12.5-inch reflector and a Schmidt camera so that I can capture closeup portraits of individual objects and image the panorama of the Milky Way."

Gregg D. Thompson is president of Dreamtech P/L, a company in Springwood, Queensland, Australia, that focuses on interior design. He is frequently active in his Springwood Observatory, which contains an 18-inch reflecting telescope. In 1990 Cambridge University Press published his *Supernova Search Charts and Handbook*, which Thompson coauthored with James T. Bryan. His sketches of deep-sky objects have appeared in many publications, including ASTRONOMY and *Deep Sky*.

The Editor

A product of Oxford, Ohio, **David J. Eicher** founded *Deep Sky Monthly* magazine at age 15 and turned it into the quarterly *Deep Sky*, one of the most respected periodicals in amateur astronomy. He served as editor-in-chief of the magazine throughout its lifetime, from 1977-1991. Eicher is associate editor of ASTRONOMY and astronomy books editor for Kalmbach Publishing Co. He has spent thousands of hours at the telescope and is well known for his sketches of astronomical objects. For his contributions to amateur astronomy the International Astronomical Union in 1990 named an asteroid in Eicher's honor, 3617 EICHER, and he has won the Caroline Herschel Astronomy Project Award (Western Amateur Astronomers) and the Lone Stargazer Award (Texas Star Party). In addition to *Stars and Galaxies*, Eicher is the author or editor of three books, *The Universe from Your Backyard* (Cambridge University Press and Kalmbach Publishing Co., 1988); *Deep-Sky Observing with Small Telescopes* (Enslow Publishers, Hillside, New Jersey, 1989); and *Beyond the Solar System* (AstroMedia, Waukesha, Wisconsin, 1992). Besides astronomy, Eicher is actively involved in Civil War history. He lives with his wife Lynda in Waukesha, Wisconsin.

Index

A

Adhara, 129
AE Aurigae, 175
Albireo, 32, 108, 111
Alcor, 50, 55, 56
Aldebaran, 166
Alkaid, 50
Alnilam, 161
Alnitak, 161
Alpha Andromedae, 58, 88, 90
Alpha Comae Berenices, 141
Alpha Pegasi, 90
Alpha Sculptoris, 76
Alpha Scuti, 116
Alpha Sextantis, 80, 83
Altair, 142
American Revolution, 35
Andromeda, 58, 76, 88, 94, 96, 179, 180
Andromeda Galaxy, 44, 45, 46, 48, 49, 94, 97, 100, 147
Anglo-Australian Telescope Board, 68, 78, 79
Ankaa, 77
Antares, 142
Aquarius, 158
Aquila, 112, 114, 182, 191, 192
Ara, 38
Aristotle, 127
Arp, Halton C., 35, 47, 85
Astrophotography, 16
Atlas of the Andromeda Galaxy, 47
Atlas of Peculiar Galaxies, 85
Auriga, 153, 166, 175
Averted vision, 65, 92, 98, 104, 111, 132, 162

B

B33, 161
B64, 147
B70, 147
B72, 147, 156
B74, 147
B85, 147
B86, 24, 26, 146, 147, 154, 155, 156
B87, 142, 147, 156
B88, 147
B89, 147
B92, 24, 25, 26, 146, 147, 154, 155, 156
B93, 24, 26, 147, 155, 156
B111, 114, 117
B119a, 114, 117
B142, 145, 147
B143, 142, 145, 147
B145, 147
B169, 147
B170, 147
B171, 147
B172, 145
B244, 147
B250,147
B289, 147
B292, 156
B300, 156
B337, 145, 147
B340, 147
B356, 147
B363, 147
Barclay, Jim, 127, 129
Barnard, Edward Emerson, 30, 114, 117, 142, 154
Barnard's Loop, 16, 18
Becvar, Antonin, 145, 155
Bedford Catalogue, 30
Beehive Cluster, 40, 137
Berkeley 58, 11
Berlin Observatory, 114
Beta Andromedae, 58
Beta Aquarii, 159
Beta Canum Venaticorum, 55
Beta Cassiopeiae, 60
Beta Ceti, 67
Beta Comae Berenices, 138
Beta Herculis, 150
Beta Lyrae, 176
Beta Pegasi, 92
Beta Ursae Majoris, 149
Betelgeuse, 105, 152
Bevis, John, 166
Big Dipper, 50, 53, 54, 84, 85, 148
Binoculars, 27, 32, 36, 44, 94, 100, 118, 123, 124, 127, 129, 187
Blinking Planetary, 191
BM Scorpii, 27, 118, 119
Bok globules, 16
Bootes, 32, 56, 138, 141
Brocchi's Cluster, 112
Brownlee, K. A., 56
Bubble Nebula, 13
Burnham's Celestial Handbook, 69
Butterfly Cluster, 27, 118

C

California Nebula, 9
Cancer, 137
Canes Venatici, 50, 53, 87, 100, 138, 140, 141
Canis Major, 102, 127, 129, 134
Capricornus, 113
Carbon, 30
Carina, 42, 121, 122, 123, 124
Carina OB1, 123
Cassiopeia, 9, 12, 58, 60, 94, 99, 107, 172
Castor, 134, 136, 137
Castor A, 134
Castor B, 134
Catalog of Peculiar Galaxies, 35
Centaurus, 36, 42, 87, 122
Centaurus A, 38
Cepheid variables, 47
Cepheus, 147, 180, 182
Cetus, 62, 66, 179, 180
Chaple, Glenn F. Jr., 136
Child, Jack, 19
Chi Ursae Majoris, 55, 87
Christmas Tree Cluster, 153
Clark, Alvan G., 30
Coal Sack, 36, 43
Collinder 121, 128, 129
Coma Berenices, 32, 35, 69, 138, 140, 141
Coma Berenices galaxy cluster, 35
Coma Berenices star cluster, 32
Coma galaxy cluster, 35
Coma-Virgo cluster of galaxies, 35, 69, 71
Comets, 69
Comet Halley, 36
Cone Nebula, 151, 153
Coombs, Lee C., 82, 105, 107, 108, 116, 120, 192
Corder, Jeffrey L., 44, 130, 182
Corona Borealis, 175
Corvus, 150, 180, 181
Crab Nebula, 166, 167, 170, 171
Crux, 42, 43, 121
Cygnus, 32, 97, 108, 111, 114, 147, 182, 191, 192

D

Dark adaptation, 65, 162
Dark nebulae, 40, 114, 142, 145, 147, 154, 156
Deep Sky magazine, 97
Delphinus, 133, 176, 179, 192
Delta Andromedae, 58
Delta Aquarii, 157
Delta Cassiopeiae, 12
Delta Ceti, 62, 98, 100
Deneb, 182
Dilsizian, Rick, 74
Dorado, 42
Double Cluster, 8, 9, 12, 30, 53, 58, 60, 71, 87, 104, 107, 108, 134
Draco, 56
Dragesco, Jean, 107
Draper, Henry, 29
Dreyer, J. L. E., 88
Dubhe, 53
Dumbbell Nebula, 178, 179, 190, 191
Dust lanes, 44

E

Eby, Tom, 33
Eclipsing binaries, 30, 137
Eicher, David J., 7, 9, 16, 50, 52, 55, 88, 96, 97, 98, 104, 105, 135, 138, 148, 149, 152, 188, 197
Eight-Burst Nebula, 181
18 Ceti, 66
84 Ursae Majoris, 56
81 Pegasi, 90
81 Ursae Majoris, 56
86 Ursae Majoris, 56
83 Ursae Majoris, 56
11 Comae Berenices, 141
Elliptical galaxies, 33, 35, 48, 80, 98, 73
Emission nebulae, 31, 40, 76, 187
Enif, 112, 113
Epsilon Cassiopeiae, 12, 174
Epsilon Cygni, 169
Epsilon Sextantis, 80
Eskimo Nebula, 179, 181
Eta Carinae Nebula, 36, 37, 40, 123
Eta Geminorum, 170
Eta Herculis, 108
Eta Orionis, 175
Eta Persei, 174
Eta Scuti, 114

F

Field, Gregory, 29, 162
58 Hydrae, 133
59 Cygni, 147
59 Leonis, 80
53 Arietis, 175
52 Cygni, 169, 186
52 Herculis, 132
Filters, 40, 151, 153
Flaming Star Nebula, 175
Floyd, Chris, 36, 40, 105, 119
Fomalhaut, 157
Fornax, 76, 99, 100
Fornax Dwarf Galaxy, 99
44 Ophiuchi, 147, 156
44 Ursae Majoris, 84
47 Tucanae, 36
42 Herculis, 132
42 Orionis, 31
4 Persei, 12
14 Vulpeculae, 191

G

G73, 49
Galactic center, 22
Galactic year, 44
Galaxies, 22, 32, 33, 34, 35, 38, 44, 47, 48, 50, 58, 60, 62, 65, 66, 67, 69, 74, 76, 77, 80, 82, 83, 84, 88, 93, 94, 100, 134, 154
Gamma Andromedae, 58
Gamma Aquiliae, 142
Gamma Cassiopeiae, 174
Gamma Comae Berenices, 33
Gamma Corvi, 150
Gamma Cygni, 182, 185
Gamma Lyrae, 111, 176
Gamma Pegasi, 88
Gamma Sagittae, 191
Gamma Sagittarii, 26, 147, 155
Gamma Sextantis, 80
Gemini, 106, 137, 179, 181
Germano, Martin C., 22, 25, 27, 33, 54, 59, 60, 67, 71, 74, 94, 99, 110, 112, 116, 133, 140, 141, 142, 150, 153, 155, 156, 174, 186
Ghost of Jupiter, 150, 181
Gleason, John P., 22, 189
Globular star clusters, 27, 36, 44, 49, 76, 97, 108, 112, 118, 130, 131, 138, 147
Goldstein, Alan, 7, 29, 32, 44, 50, 69, 118, 134, 176, 182, 188, 196
Gordon, Dan, 19
Great Galactic Dark Horse, 145
Great Rift, 97
Great Square of Pegasus, 58, 88, 93

H

Hallas, Tony, 16, 23, 26, 34, 35, 45, 51, 57, 95, 96, 152, 177, 183
Hamler, Walter E., 185
Harrington, Phil, 7, 22, 30, 114, 127, 172, 196
Harvard 20, 111
Helical Nebula, 157, 158
Helium, 30
Hercules, 27, 109, 127, 150, 179, 180
Hercules Cluster, 108, 111, 113
Herschel, John, 29, 30
Herschel, William, 25, 29, 114, 142, 159, 169
Hicks, Scott, 77, 110, 146, 169, 186
Higgins, David, 7, 62, 80, 88, 130, 138, 151, 154, 157, 161, 196
Hodge, Paul, 47
Homunculus Nebula, 40
Hooley, Mace, 52, 65, 104, 105, 153, 173
Horsehead Nebula, 16, 19, 160, 161, 162, 164, 165
Hubble, Edwin, 47, 69
Hug, Gary, 86
Huygens, Christiaan, 29, 30
Hyades, 32, 166
Hydra, 80, 133, 181
Hydrogen, 30, 161

I

Iburg, Bill, 46, 61, 92, 97, 99, 101, 113, 159
IC 10, 11, 58, 59, 60
IC 59, 174, 175
IC 63, 11, 174, 175
IC 289, 11
IC 351, 11
IC 405, 174, 175
IC 410, 153, 174, 175

IC 434, 16, 19, 161, 164
IC 443, 170
IC 749, 87
IC 750, 87
IC 1276, 131, 132
IC 1287, 116, 117
IC 1295, 116, 117
IC 1318, 185
IC 1805, 9, 11, 12, 60, 172, 173, 174
IC 1848, 11, 12, 60, 172, 173, 174, 175
IC 2003, 11
IC 2391, 40, 123
IC 2602, 40, 42, 122, 123
IC 5067-70, 186
IC 5068, 185
Interacting galaxies, 85
Iota Hydrae, 83
Iota Orionis, 19, 31
Iota Aquilae, 114

J

Jakiel, Richard W., 7, 84, 87, 196
Jamison, D. B., 106
Jewel Box Cluster, 3, 121, 123, 124, 125
Jones, Jeffrey L., 65, 81
Justis, Preston Scott, 46, 83, 85, 86, 132, 139

K

Keyhole Nebula, 40
Keystone, 108
Kirch, Gottfried, 114

L

Lagoon Nebula, 22, 23, 142, 146, 147
Lambda Aquilae, 114
Lambda Canis Majoris, 129
Lambda Orionis Nebula, 16, 18
Lambda Sagittarii, 27, 113
Lambda Scorpii, 26
$Lambda^1$ Sculptoris, 79
$Lambda^2$ Sculptoris, 79
Large Magellanic Cloud, 36, 41, 42, 77
LDN 1470, 11
LDN 1471, 11
Lenticular galaxies, 35
Leo, 32, 80
Leo I, 59, 60
Libra, 130
Lick Observatory, 59
Liddell, Joe, 35, 90, 181
Lilge, Alfred, 118, 145
Ling, Alister, 7, 102, 196
Little Dumbbell Nebula, 12, 13
LMC, 42
Local Group of galaxies, 44, 60, 84, 99, 100
Lucas, Steve, 7, 58, 196
Luginbuhl, Christian, 7
Lyra, 105, 108, 111, 176, 188

M

M Centauri, 38
M1, 166, 170
M2, 159
M3, 138, 139, 141
M6, 22, 24, 27, 118, 119, 120
M7, 22, 24, 26, 27, 118, 119, 120
M8, 22, 24, 110
M9, 147
M10, 111, 112, 113
M11, 114, 117, 129
M12, 111, 112, 113
M13, 27, 108, 109, 111, 112
M15, 112, 113
M20, 24, 25, 110, 147
M21, 24, 25
M22, 22, 24
M22, 27, 110, 112, 113
M24, 24, 25, 146, 154, 155
M25, 119, 120, 116, 117
M27, 12, 112, 178, 179, 181, 190, 191, 192
M31, 44, 48, 58, 76, 94, 95, 97, 100
M32, 44, 46, 48, 49, 94, 98, 100
M33, 77, 79, 96, 99, 100
M34, 11
M35, 105, 106, 107, 107
M36, 106, 107, 153
M37, 106, 107
M38, 106, 107
M41, 102, 103, 104, 105, 106, 107, 126, 127, 128
M42, 16, 29, 29, 30, 40, 137
M43, 19, 29, 30
M44, 137
M46, 104, 105, 107
M47, 104, 105, 107
M49, 71, 73, 73
M51, 50, 50, 51, 54, 56
M53, 140, 141
M55, 110, 112, 113
M56, 108, 111, 112
M57, 108, 176, 179, 181, 188, 192
M58, 71, 74, 140
M59, 71, 74, 140
M60, 71, 74, 140
M61, 71, 73
M63, 53, 54
M64, 97, 100
M75, 110, 112, 113
M76, 11, 12
M77, 62, 63, 65, 66, 98, 100
M78, 31
M81, 52, 53, 54, 84
M82, 52, 53, 54, 84
M84, 71
M86, 71, 73
M87, 71, 73, 74, 75
M89, 71, 73, 140
M90, 71, 73, 74
M93, 102, 104, 105, 107
M94, 53, 54, 55
M97, 84, 148, 149, 150
M101, 50, 54, 55, 56, 84
M102, 54, 56
M103, 11, 12
M104, 76
M106, 53, 54, 55, 87
M107, 111, 112, 113
M108, 54, 84, 86, 149
M109, 54, 56, 84
M110, 48, 99
Maffei, Paolo, 60
Maffei 1, 60
Magellanic Clouds, 36
Maley, Paul, 103
Marcus, Dan, 159
Marling, Jack B., 31, 38, 53, 59, 104, 142, 146, 149, 150, 160, 176, 178, 179, 180, 182, 190, 191
Mayer, Ben, 120
Mayerchek, Mike, 27, 120
McDonald, Bruce D., 49
Melotte 15, 172
Melotte 20, 11
Melotte 111, 35
Merak, 54, 84, 148
Messier, Charles, 30, 48, 69, 99, 100, 114, 117, 142, 159
Metcalfe, Jack, 133
Milky Way, 7, 9, 16, 22, 25, 26, 36, 44, 60, 69, 76, 88, 97, 100, 102, 108, 113, 114, 117, 118, 127, 130, 141, 142, 147, 151, 154, 155, 161, 181, 182
Mintaka, 161
Mizar, 50, 55, 56
Monoceros, 151, 152
Moon, 9, 44, 100, 102, 104, 105, 106, 116, 121, 123, 127, 153, 186
Mount, Daphne, 16, 23, 26, 34, 35, 45, 51, 57, 95, 96, 152, 177 183
Mu Cephei, 191
Mu Columbae, 175
Mu Hydrae, 150
Mulitple stars, 32
Mullaney, James, 136
Murzam, 128
Musca, 38

N

National Optical Astronomy Observatories, 125
Nebula filters, 67, 148, 157, 158
Nebulae, 30, 36, 114, 134, 150, 151, 153, 161, 174
Nebulosa, 94
Nelson, Brant, 185
Neon, 30
Neptune, 181
New General Catalogue, 88
Newton, Jack, 39, 47, 53, 63, 71, 73, 75, 98, 102, 104, 109, 110, 111, 112, 140, 142, 149, 150, 151, 158, 171, 172, 180, 186, 192
NGC 1, 88, 90, 91, 93
NGC 2, 88, 90, 91, 93
NGC 16, 90, 91, 93
NGC 18, 179
NGC 23, 90, 93
NGC 26, 90, 93
NGC 40, 180, 181
NGC 55, 76, 77, 78, 79
NGC 103, 11
NGC 104, 42
NGC 129, 11
NGC 136, 11
NGC 147, 11, 47, 49, 94, 99, 100
NGC 175, 66, 67
NGC 178, 66
NGC 185, 47, 49, 94, 99, 100
NGC 205, 44, 46, 48, 49, 94, 99, 100, 100
NGC 206, 46, 47, 97
NGC 210, 66
NGC 246, 64, 65, 66, 67, 179, 180, 181
NGC 247, 66, 67, 76, 78, 79
NGC 253, 76, 77, 78, 99, 100, 101
NGC 255, 65, 66, 179
NGC 273, 66, 67
NGC 274, 66, 67
NGC 275, 67
NGC 281, 13,
NGC 288, 76, 78, 99
NGC 300, 76, 77, 78, 79
NGC 346, 36
NGC 362, 36, 42
NGC 371, 36
NGC 457, 11, 12, 107
NGC 559, 11, 12
NGC 604, 100
NGC 663, 11, 12
NGC 744, 11
NGC 869, 9, 11, 107
NGC 884, 9, 11, 107
NGC 891, 58, 59, 60, 61, 76
NGC 957, 11
NGC 1003, 11
NGC 1023, 11
NGC 1027, 11
NGC 1055, 65, 66
NGC 1058, 11
NGC 1097, 99
NGC 1245, 11
NGC 1333, 11
NGC 1342, 11
NGC 1491, 11
NGC 1499, 11
NGC 1513, 11
NGC 1528, 11
NGC 1545, 11
NGC 1579, 11, 174, 175
NGC 1760-9, 36
NGC 1893, 153
NGC 1907, 106, 107
NGC 1931, 153, 174, 175
NGC 1973-5-7, 30, 31
NGC 1980, 31
NGC 1981, 31
NGC 2023, 19, 161, 162
NGC 2024, 16, 19, 161
NGC 2032-5, 36
NGC 2070, 42
NGC 2158, 105, 106, 107
NGC 2204, 128
NGC 2237-9, 153
NGC 2239, 151
NGC 2243, 127, 128, 129
NGC 2244, 16, 105, 106, 107, 151
NGC 2246, 151
NGC 2264, 151, 153
NGC 2354, 127, 128, 129
NGC 2360, 126, 128, 129
NGC 2362, 126, 128
NGC 2392, 179, 181
NGC 2438, 104, 107
NGC 2516, 40, 42, 123, 124
NGC 2808, 38
NGC 2967, 82, 83
NGC 2974, 82, 83
NGC 2990, 80, 82
NGC 3044, 82, 83
NGC 3055, 82, 83
NGC 3114, 40, 124
NGC 3115, 80, 81, 82, 87
NGC 3132, 148, 150, 178, 181
NGC 3156, 80, 82
NGC 3165, 80, 82
NGC 3166, 80, 82, 83
NGC 3169, 80, 82
NGC 3242, 148, 150, 181
NGC 3250, 178
NGC 3293, 42, 121, 123
NGC 3372, 42
NGC 3423, 80, 82
NGC 3448, 84
NGC 3532, 40, 42, 124
NGC 3594, 84
NGC 3631, 84, 86
NGC 3675, 85
NGC 3718, 85
NGC 3726, 85, 86
NGC 3729, 85, 87
NGC 3766, 42, 122, 123
NGC 3877, 85, 87
NGC 3938, 87
NGC 3953, 87
NGC 4013, 87
NGC 4026, 87
NGC 4051, 87
NGC 4088, 87
NGC 4096, 87
NGC 4100, 87
NGC 4144, 87
NGC 4147, 141
NGC 4151, 87
NGC 4216, 69, 71
NGC 4324, 71
NGC 4361, 148, 150, 180, 181
NGC 4402, 71, 73
NGC 4413, 71, 73
NGC 4425, 71, 73
NGC 4435, 71, 73
NGC 4438, 71, 73
NGC 4448, 33, 35
NGC 4485, 53, 54, 55
NGC 4490, 53, 54, 55
NGC 4494, 33, 35
NGC 4526, 71, 73
NGC 4559, 33, 34, 35
NGC 4565, 32, 33, 35, 69, 76
NGC 4567, 71, 74
NGC 4568, 71, 74

NGC 4638, 71, 74
NGC 4647, 71, 74
NGC 4712, 34, 35, 35
NGC 4725, 32, 34, 35, 35
NGC 4747, 35, 35
NGC 4755, 42, 121
NGC 4833, 38
NGC 4874, 35
NGC 4889, 35, 35
NGC 4945, 38
NGC 5053, 140, 141
NGC 5128, 38, 38, 42, 87
NGC 5139, 42
NGC 5195, 50, 54
NGC 5286, 38
NGC 5466, 140, 141
NGC 5694, 131, 133
NGC 5897, 130, 131, 132, 133
NGC 5907, 54, 56, 76
NGC 6067, 123, 124
NGC 6181, 180
NGC 6210, 148, 150, 179, 180, 181
NGC 6229, 131, 132, 133
NGC 6231, 24, 27, 120
NGC 6397, 38, 42
NGC 6453, 118
NGC 6517, 131, 132
NGC 6520, 24, 26, 27, 147, 154, 155
NGC 6530, 25
NGC 6539, 130, 131, 132
NGC 6603, 24, 25
NGC 6649, 116, 117
NGC 6664, 116, 117
NGC 6704, 116, 117
NGC 6712, 116, 117
NGC 6744, 38
NGC 6752, 38
NGC 6781, 191, 192
NGC 6826, 191, 192
NGC 6888, 169, 170, 186
NGC 6905, 176, 179, 180, 181, 190, 192
NGC 6914, 186
NGC 6960, 168, 169, 170, 186
NGC 6979, 169, 186
NGC 6992, 186
NGC 6992-5, 168, 169, 170, 186, 187
NGC 6995, 186
NGC 7000, 182, 186
NGC 7006, 131, 132, 133
NGC 7008, 191, 192
NGC 7009, 159
NGC 7293, 159
NGC 7448, 90, 93
NGC 7463, 92, 93
NGC 7464, 92, 93
NGC 7465, 92, 93
NGC 7640, 180
NGC 7662, 179, 180, 181
NGC 7673, 92, 93
NGC 7677, 92, 93
NGC 7678, 92, 93
NGC 7769, 90, 93
NGC 7770, 90, 93
NGC 7771, 90, 93
NGC 7793, 76
NGC 7814, 88, 89, 92, 93
9 Bootis, 138
19 Ceti, 66, 67
93 Leonis, 141
Nitrogen, 30
Norma, 123
North America Nebula, 182, 183
Novae, 47
Nu Aquarii, 159

O

OB associations, 85
Omega Centauri, 36, 39

Omega^2 Cygni, 191
Omicron^1 Canis Majoris, 129
Omicron Cassiopeiae, 49
Omicron Leonis, 83
Open Star clusters, 31, 102, 117, 118, 127, 159
Ophiuchus, 97, 111, 113, 132, 147, 156
Orion, 16, 18, 29, 31, 40, 105, 127, 137, 152, 175
Orion's belt, 18, 161
Orion Nebula, 12, 16, 19, 29, 31, 100, 137, 147, 161
Owl Cluster, 107
Owl Nebula, 84, 148, 149, 150
Oxygen, 30, 176

P

Palomar Mountain Observatory, 131
Parrot's Head Nebula, 142, 147, 156
Parsons, William, 159
Pavo, 38
Pegasus, 44, 88, 112, 113
Pelican Nebula, 185
Perseus, 9, 166
Phecda, 54, 87
Phi Aurigae, 175
Phi Cassiopeiae, 107
Phi Persei, 12
Phoenix, 77
Photographic Atlas of Selected Regions of the Milky Way, A, 142
Pinwheel Galaxy, 99
Pisces, 96, 97
Piscis Austrinus, 157
Planetary nebulae, 117, 148, 157, 176, 179, 180, 192
Pleiades, 32
Pommier, Rod, 7, 76, 197
Potter, Ron, 59
Proboscis Major, 30
Proboscis Minor, 30
Puppis, 104

Q

q Velorum, 148
Quirk, Steve, 76

R

R Scuti, 116, 117
Radloff, Max, 7, 94, 197
Realm of the galaxies, 69
Reeves, Robert, 19
Reflection nebulae, 161
Regio Huygeniana, 30
Regio Messieriana, 30
Regulus, 60
Reticulum, 42
Rho Ophiuchi, 142
Rich-field telescopes, 25
Rigel, 134, 136, 137, 161
Rigel A, 134
Rigel B, 134
Ring Nebula, 105, 176, 177, 179, 181, 188, 189
Roques, Paul, 48, 55, 154, 156
Rosette Nebula, 16, 18, 105, 106, 151, 152
Rotramel, Rick, 168
Royer, Ronald, 37, 41, 43, 115, 145, 187
Runyan, Chesley Jr., 157
RV Tauri, 116

S

S Andromedae, 47

S Monocerotis, 151
S Nebula, 145, 147, 156
Sagitta, 112, 190
Sagittarius, 22, 26, 110, 113, 114, 119, 142, 147, 154, 182
Sagittarius star cloud, 147
Saturn, 159
Saturn Nebula, 158, 159
Schrantz, Rick, 32
Schur, Chris, 7, 18, 108, 166, 168, 170
Scorpius, 22, 26, 27, 77, 118, 119, 147
Sculptor, 76, 78, 99
Sculptor group of galaxies, 75, 100
Scutum, 106, 117, 129
Scutum star cloud, 25, 114, 115
Seagull Nebula, 36
Serpens Cauda, 130
17 Ceti, 66
17 Comae Berenices, 32, 33, 35
17 Virginis, 71
Sextans, 80, 82, 83, 87
Seyfert, Carl, 100
Seyfert galaxies, 62, 87, 100
Siamese Twins, 74
Simeis 147, 170
Sinus Magnes, 30
Sirius, 102, 127, 128, 129, 134, 135, 137
Sirius A, 134
Sirius B, 134
Sisk, Mike, 29, 94
16 Cygni, 191
60 Cygni, 147
64 Ophiuchi, 132
69 Ophiuchi, 132
67 Ursae Majoris, 87
Skiff, Brian A., 7
Sky Catalogue 2000.0, 69
Sliva, Roger, 94
Small Magellanic Cloud, 36, 41, 42
Small Sagittarius star cloud, 25, 26
Smith, Bill, 159
Smyth, William Henry, 30, 31, 114
Snake Nebula, 156
Sorrells, Bill, 91
South Galactic Pole, 78
Southern Pleiades, 123
Spiral arms, 53, 100
Spiral structure, 80, 96, 87
Star clouds, 44, 97
Star clusters, 25, 36, 102, 114, 147, 151
Stecker, Michael, 165
Stock 2, 11
Struve, F. G. W., 30
Struve 743, 31
Struve 750, 31
Struve 2325, 116
Sulfur, 30
Supergiant stars, 123
Supernovae, 47, 53
Supernova remnants, 166, 170
Surface brightness, 12, 40, 48, 49, 54, 58, 59, 67, 73, 74, 80, 83, 87, 92, 98, 117, 130, 132, 140, 150, 153
Sword of Orion, 18

T

Tarantula Nebula, 36, 40
Tau Canis Majoris, 128
Taurus, 32, 166
Theta Ophiuchi, 156
Theta Orionis, 29
Theta^1 Orionis, 30
Theta^2 Orionis, 30
30 Ophiuchus, 113
35 Librae, 130

31 Comae Berenices, 34
Thompson, Gregg D., 7, 36, 38, 121, 197
Trapezium, 29, 30
Triangulum, 77, 96
Trifid Nebula, 25, 26, 147
Triple stars, 32
Trumpler 2, 11
Trumpler, Robert J., 142
Tucana, 42
TV Cassiopeiae, 60
12 Comae Berenices, 32, 35
12 Monocerotis, 151
20 Librae, 130, 133
25 Sextantis, 83
24 Canum Venaticorum, 50
24 Comae Berenices, 32, 35
24 Ursae Majoris, 53
23 Ursae Majoris, 53
12 Delphini, 133
2 Cygni, 108
2 Sextantis, 83

U

University of Chicago, 142
Upsilon Aquarii, 157
Upsilon Pegasi, 92
Uranus, 181
Ursa Major, 52, 53, 54, 55, 84, 87, 100, 148, 149
Ursa Major moving group, 32

V

Variable stars, 27, 30, 116, 175, 176
Vaucouleurs, Gerard de, 77
Vaughn, Chuck, 175
Vega, 176, 188
Veil Nebula, 168, 169, 170, 186, 187
Vela, 42, 123, 148, 178, 181
Virgo, 69, 74, 140
Virgo A, 73
Virgo Cluster, 48, 68, 69, 72, 100, 141
Virgo Supercluster, 84
Viscome, George, 24, 114, 126
Vulpecula, 12, 112, 178, 179, 190

W

Webb, Thomas W., 31
Wesen, 127
Whirlpool Galaxy, 50, 57
Wild Duck Cluster, 106, 114
Wolf, Max, 182
Wolf-Rayet stars, 186

X

X Sagittarii, 22
Xi Puppis, 104

Y

YY Geminorum, 137

Z

Zeta Cancri, 136, 137
Zeta Ophiuchi, 111, 113
Zeta Orionis, 16, 19
Zeta Persei, 175
Zeta Sagittae, 111
Zeta Sagittarii, 113
Zeta Tauri, 166
Zussman, Kim, 52, 53, 56, 62, 64, 72, 76, 89, 167, 178, 179